CNC Programming Using Fanuc Custom Macro B

S. K. Sinha

New York   Chicago   San Francisco
Lisbon   London   Madrid   Mexico City
Milan   New Delhi   San Juan
Seoul   Singapore   Sydney   Toronto

The McGraw·Hill Companies

Library of Congress Cataloging-in-Publication Data

Sinha, S. K. (Sanjay Kumar), date.
  CNC programming using Fanuc custom B macro/S.K. Sinha.
      p.   cm.
  Includes index.
  ISBN 978-0-07-171332-0 (alk. paper)
  1. Machine-tools—Numerical control Programming.   2. Macro instructions
(Electronic computers)   I. Title.
  TJ1189.S5686   2010
  621.9′023—dc22   2010015377

McGraw-Hill books are available at special quantity discounts to use as premiums and sales promotions, or for use in corporate training programs. To contact a representative please e-mail us at bulksales@mcgraw-hill.com.

CNC Programming Using Fanuc Custom Macro B

1 2 3 4 5 6 7 8 9 0 DOC/DOC 1 9 8 7 6 5 4 3 2 1 0

ISBN 978-0-07-171332-0
MHID 0-07-171332-8

Sponsoring Editor
  Michael Penn

Acquisitions Coordinator
  Michael Mulcahy

Editorial Supervisor
  David E. Fogarty

Project Manager
  Harleen Chopra,
  Glyph International

Copy Editor
  Ragini Pandey,
  Glyph International

Proofreader
  Christine Andreasen

Indexer
  BIM Indexing &
  Proofreading Services

Production Supervisor
  Richard C. Ruzycka

Composition
  Glyph International

Art Director, Cover
  Jeff Weeks

The pages within this book were printed on acid-free paper.

Dedicated
to
the everlasting memory of my father
Dr. Rameshwar Prasad Sinha
who inculcated engineering traits in me

About the Author

S. K. Sinha earned his Ph.D. in mechanical engineering from Indian Institute of Technology, Kanpur in 1993. He has 20 years of teaching/industrial experience and has been working in the CNC area for the past 12 years. He has also authored a book on basic programming techniques for Fanuc 0i series controls, *CNC Programming*, 8th edition, published by Galgotia Publications Pvt. Ltd., New Delhi, India.

Contents

Preface

Due to excellent accuracy, repeatability, and several other unparalleled and exotic features, computer numerical control (CNC) machines have now virtually completely replaced manual machines in the manufacturing industry. Even small manufacturing units are now using CNC machines because these prove to be cheaper in the long run. In fact, "why CNC?" is not at all a topic of discussion these days!

CNC machines with recently introduced *macro programming* features have become so powerful that not using the macro features is like using the latest mobile handset only for exchanging voice messages. Unfortunately, there is a lack of adequately trained personnel in this area. The main reason is that because this is an industry-oriented subject, engineering colleges/polytechnics do not deal with it in sufficient detail. Even the commercial CNC training schools generally provide only basic CNC training, and the few training school that do provide advanced training charge exorbitantly. The worst part is that, because of professional competition, people are generally unwilling to share their knowledge with their colleagues; in such a scenario, it is difficult to learn macro programming, which is often rightly described as *the best kept secret of CNC*.

The root cause of hindrance in learning macro programming is the lack of suitable books in this area. Even the manuals which come with the machines do not serve the purpose because, though these do describe all the features of the language, these are actually reference books not written in a textbook style. As a result, while these are good for getting detailed information about some specific programming feature, one cannot really learn the language in a systematic manner from them. What is required is a text with step-by-step instructions, starting from the very basic principles and gradually proceeding further in order of complexity. This was the prime motivation for the present book. It is a self-sufficient text, designed to be read from beginning until the end, in chapter sequence. No external help, in the form of any other book or an instructor, would be needed. One can learn simply by *self-study*.

The present text specifically deals with the Fanuc version of macro programming language called *Custom Macro B* (*Custom Macro A* is outdated and no longer used), with reference to 0i series controls. Though the language is the *same* on all control versions of Fanuc (as well as on Fanuc-compatible controls), some of the *system variables* and *control parameters* differ on different control versions. Therefore, one would need to verify these from the respective machine manuals and make necessary changes wherever required.

Finally, note that though every care has been taken to ensure that the programs given in the text are error free and work as intended, neither the author nor the publisher assumes any responsibility for any inadvertent error that might have crept in. As a rule of thumb, one should always check the simulation of a new program *before* actual execution.

Suggestions for further improvement would be gratefully acknowledged. Any feedback can be sent directly to the author at sinha_nsit@yahoo.co.uk.

S. K. Sinha

Acknowledgments

The author is extremely thankful to Prof. V. P. Singh, Professor-in-charge, Main Workshop, Institute of Technology, BHU, Varanasi, India, and Mr. M. D. Tyagi, Workshop Superintendent, for allowing free access to the CNC Laboratory, where most of the programs given in the book were tested. Thanks are also due to Prof. J. P. Dwivedi, HOD, Mechanical Engineering Department, and my colleague Dr. Pradyumna Ghosh for their words of encouragement.

The author has also used some of the suggestions of professional CNC experts such as Mr. Dan Fritz, a former Fanuc Applications Engineer and presently President, Suburban Machinery Software, Inc., and Mr. Steve Hase , a CNC wizard from Waukesha, Wisconsin. Sincere thanks are due to them.

The entire editorial/production team of McGraw-Hill, and Glyph International, where typesetting was done, are to be thanked for their professional help with this project.

Last, but not the least, the author is thankful to his wife Seema, daughter Shubhi, and son Sarthak, who very patiently forbore his long working hours during the course of this work.

Finally, the author would be failing in his duty if he does not mention the name of his mother, Mrs. Kanti Singh, for being a constant source of encouragement for him.

S. K. Sinha

CHAPTER 1
Introduction

Numerical control is an application of digital technology to control a machine tool by actuating various drive motors, relays, etc. by a series of coded instructions, called *part programs*. These machines were initially called *numerically controlled* (NC) machines. NC technology was made commercially available in mid-1950s. Later, as the technology advanced, a computer was made an integral part of the machine, and most of the functions of the machine were controlled through software rather than dedicated hardware for each function. Such machines were called *computer numerical control* (CNC) machines. NC machines are not manufactured anymore, though we still use this term for referring to this technology.

Due to excellent accuracy, repeatability, and several other unparalleled and exotic features, the CNC machines have now virtually completely replaced the manual machines in the manufacturing industry. Even small manufacturing units are now using CNC machines, because these prove to be cheaper in the long run. In fact, "why CNC?" is not at all a topic of discussion these days.

However, as with any other computer application, it is imperative to develop good and efficient programs to exploit the full potential of CNC machines. Unfortunately, there is a lack of adequately trained manpower in this area. The main reason is that the age-old curriculum of engineering colleges/polytechnics does not deal with CNC programming in detail. Moreover, if you are a practicing engineer, you may have realized that the few people in industry who indeed have good programming knowledge are not too willing to share their knowledge and experience with others, for obvious reasons! And, learning on one's own is not really so easy, particularly if the subject is complex and the information is available in a scattered manner.

To cater to this need, a number of commercial CNC training schools have come up recently. However, most of these provide only basic CNC training. Advanced CNC training is provided by very few training schools in the whole world. Even the relevant books, presented in a self-contained textbook style, are not available. And, manuals/handbooks are mainly useful for reference only, when one wants to see a detailed description of some specific feature. This makes learning by self-study difficult, especially for a newcomer who does not even know how to start. In fact, if you want to master a complex subject, to which you are completely new, on your own, the first thing you need to know is **how to start learning**. And, if you fail to choose the right track, very soon the subject will appear frustratingly complex and boring. So, a methodical and step-by-step approach is extremely important.

This book aims at fulfilling this need. If you have some basic knowledge of programming and operation of CNC machines, this book will help you in learning the advanced features of CNC machines, features you may not even have heard of. And, the most important thing is that, the book is designed in such a manner that **you will learn on your own**!

We will first discuss the various part programming techniques, and their areas of application. This will be followed by a discussion of an advanced programming technique which adds a whole new dimension to the conventional CNC programming. The remaining chapters of the book are devoted to a detailed description of this technique.

1.1   Part Programming Techniques

A CNC machine is only as good as the part programs being used for machining. The role of the machine operator is generally limited to clamping and unclamping the workpiece, for which even a semi-skilled worker would be good enough. The quality and efficiency of production depends mainly on the programming skill of the programmer. The programmer need not be an expert machinist, but he must have a very good knowledge of shop floor practices. In addition, he must know the available control features of his machine. It is only then, efficient part programs, producing good results, can be developed. A CNC machine works just like an "obedient slave," **doing exactly what it is instructed to do, in a very precise manner**. Hence, a CNC machine would be worth its name, only if the instructions given to it, through the part programs, offer the best possible solution to perform a given task.

Broadly speaking, there are four different ways of preparing part programs, described in brief in the following sections.

Conventional Part Programming

Conventional part programming, which is a simple G-code/M-code programming, suffers from several limitations and has limited scope. If one needs to machine only limited types of workpieces in a routine manner, without bothering much about the efficiency of the production activity, conventional part programming would generally be adequate. The limitation of conventional programming is that it does not allow the use of variables, mathematical operations, functions, logical statements, looping, and the like in a program. In other words, it does not have even the basic features of a typical computer language such as PASCAL. It is just a rigid type program **for a particular requirement**. This obviously limits the scope of programs to a great extent. It is not possible to write an "intelligent" part program with built-in logic for different requirements. In fact, as we will see later, there are a number of applications where conventional part programming cannot be used. It only serves the limited purpose for which it was designed.

Conversational Part Programming

Even conventional part programming is fairly complex. So, for the purpose of simplifying programming for certain common applications, *conversational* or *lead-through* programming (referred to as *blue print programming* or *direct dimension programming*) was introduced, which enables users to even without having adequate programming experience, easily develop suitable part programs. The programmer need not know the part programming language in detail; he only has to know what is to be done in what sequence, and with what cutting parameters. The control prompts him for all the required information, in an **interactive manner**. This method of programming, however, suffers from the inherent limitation of being applicable only to certain specific geometries. So, even though it is possible to quickly write efficient programs for some common applications, this method is useless for special requirements. In reality, conversational programming is only a small subset of what can be done with conventional part programming technique. In fact, conversational programming is not meant for qualified engineers.

Part Programming Using CAM Software

As long as only two axes of a machine are required to be **simultaneously** controlled (including helical interpolation), which is usually the case in most of the practical applications, it is possible to write suitable part programs manually, unless the geometry is nonstandard, for example, involving a parabolic segment. For nonstandard and/or three-dimensional geometries, such as die machining, manual development of part programs might become too tedious or even impossible in certain cases. Several CAM softwares have been developed, which generate the required toolpath for the desired machining, and automatically prepare a part program to suit the selected control version. However, the basic purpose of CAM software is **to calculate what cannot be calculated manually** (e.g., the toolpath for a 5-axis machining where it is desirable to always keep the axis of the tool perpendicular to the surface it is machining). So, even though it can generate a complex toolpath, and give certain machining-related information such as machining time, interference check, etc., this is all it can do. Broadly speaking, it does not add additional features to conventional part programming. The difference between conventional part programming and computer-aided part programming is very much like the difference between manual arithmetic calculations and calculations using a calculator. So, CAM software is useful only for the purpose for which it is designed. It does not offer total manufacturing solution. There are requirements, as we will see later, for which it becomes simply useless.

Macro Programming

In the 1990s, basic features of high-level computer languages were incorporated in the conventional part programming language. The new programming language was called *macro programming*, the features of which closely resemble the BASIC computer language. Over the years, macro programming has seen steady development. It is now quite advanced, except that alphanumeric characters still cannot be used as variable names, among certain other limitations. Macro programming, as we will see later, has completely changed the way CNC machines are programmed these days, opening up endless possibilities, limited only by the imagination of the programmer.

Comparison among the Four Methods

Each out of the four ways to prepare part programs has its own area of application, serving its specific purpose. Any comparison among these is thus meaningless. Macro programming, however, offers several tools for enhancing productivity, which is not possible with any of the other three methods. The emphasis today is not just on automation, to boost productivity; it is also on **flexible automation**, so as to respond quickly to the fast-changing market requirements. Macro programming can make the part programs so flexible with built-in logic that the same program can be used for different machining requirements. Moreover, in many cases, a CNC machine can be programmed to make its own decision in different situations, without any intervention from the operator. All this results in unparalleled productivity enhancement.

1.2   Certain Applications of Macro Programming

Many manufacturing units have excellent applications for macro programming, but they are not even aware of this fact. And, if one does not know that there is an application for something, one will never consider learning it, let alone using it! In fact, most of the modern CNC machines have this capability as a built-in feature (optional on some machines), but many users do not even know this. In today's world of cut-throat competition, one cannot really afford to be totally ignorant of such a powerful feature of CNC machines. Given the enhancements that this kind of programming brings, it is surprising that most of the *machine tool builders* (MTB), control manufacturers, and even commercial training schools do not say much about it.

The major application areas of macro programming are discussed in the following sections.

Complex Motions

Most CNC controls provide only linear and circular motions. As a result, it is not possible, for example, to do parabolic turning on a lathe or to cut an elliptical groove on a milling machine. For such applications, one has to take help of some external computer programming language for generating the toolpath, as a chain of very small straight line segments, joined sequentially by linear interpolation (G01). The other way would be to use CAM software. However, macro programming, which has all the relevant features of a high-level computer programming language, enables us to easily generate any type of complex motion for which mathematical equations are available. It also obviates the problem of loading into the CNC's RAM, the extremely large file generated by the external computer program or the CAM software, which sometimes makes macro programming a much better option. While macro programming cannot completely replace CAM software, it can certainly do many things that may not be possible without the use of CAM software.

Families of Parts

Almost all companies have some applications that fit into this category. Part programming time can be reduced by developing *parametric programs* for *families of parts*. An example is bolt-hole drilling on a flange. While all the flanges may belong to the same family by virtue of similarity in design and production method, they may all be actually different. For example, the number of holes, the depth of holes (i.e., the thickness of the flange), and the diameter of the pitch circle may be different in different flanges. Even the workpiece materials may be different, requiring different feedrates and rpms. These differences may also necessitate the use of different machining cycles (G81, G82, etc.) in different flanges.

This means that even though the machining methods for all the flanges are essentially the same, all the programs would be somewhat different. However, a **single** program, using the macro programming features, would do the job. We only have to identify the varying entities and write a program with these entities as variables, defined in the beginning of the program. It is also possible to define a subroutine, with the variable quantities as its *arguments*. For a particular flange, the *main program* (i.e., the program at the top-most level in nested programs) can simply call the subroutine, with the desired values for its arguments. The subroutine would be executed with the specified values for the variable quantities. The programming methodology is very similar to that in any high-level computer language. The major difference lies in the way the variables are designated.

Note that a subroutine called without an argument list is called a *subprogram*, whereas if it is called with an argument list, it is called a *macro program* (or, simply a *macro*). A built-in macro, which is made available on the machine by the machine tool builder, is called a *machine macro*, and a user-defined macro is referred to as a *custom macro*. Subprograms are called by M98 (which does not permit an argument list), whereas macros are called by G65 or G66 (which permit an argument list). Note that G65/G66 need not necessarily have an argument list (though this is a rare possibility). A macro is designed to be called with an argument list. A subprogram, on the other hand, is designed to be called without an argument list. The same program behaves like a subprogram if it is called by M98, whereas it behaves like a macro if called by G65 or G66 (with or without an argument list). The difference between a subprogram and a macro is explained in detail in Chaps. 6 and 7.

Using a macro for a family of parts is often referred to as *parametric programming*; the arguments of the macro are the *parameters* that define the specific function of the macro. Here, the term "parameter" is used in the **mathematical sense**. It is not even remotely related to *control parameters*, which define the default settings of the machine. Control parameters are also referred to as system parameters, CNC/PMC parameters, machine parameters, or, simply, parameters.

It is also possible to define a new G-code (say, G100) or a new M-code (say, M100), corresponding to a macro. Then, such a macro can also be called by the defined G- or M-codes. In fact, **the new codes become available to all the programs on that particular machine, and all the programmers can use them**. This simplifies programming as the programmers need not even know how to call a macro; they only need to know how to use the new codes that become similar to the ordinary G-codes/M-codes. Of course, an expert programmer has to develop the new codes initially.

Custom Canned Cycles

CNC machines are equipped with a large number of *canned cycles* for different machining requirements, which simplify programming to a great extent. For example, if one has to generate G71 (multiple turning cycle) type toolpath using G00 and G01, it would be prohibitively complex and lengthy because of enormous number of mathematical calculations involved. The same exercise would have to be repeated for every job with a different geometry. But, G71 is just a two-line code, for all geometries.

However, a situation may arise when no predefined canned cycle would be quite suitable for a particular application. For example, if a very deep hole is to be drilled, it may be desirable to progressively reduce the peck length in subsequent pecks, to allow the chips to come out of the hole easily, especially when the tool is deep inside the hole. Such a canned cycle is not available on any CNC machine. All the peck-drilling cycles use uniform peck length throughout the depth of the hole (except the last peck, which is adjusted to suit the hole depth). However, using the macro programming technique, it is possible to write a program with varying peck length (as per user-defined logic), and as mentioned earlier, it is also possible to define a new G-code for such a program, which can, then, be very conveniently used the way a built-in canned cycle is used. Development of new canned cycles customizes CNC machines, as per individual needs.

"Intelligent" Programs

Macro programming allows the use of conditional statements (IF_ THEN,) to be used with conditional operators (*equal to, not equal to, less than, less than or equal to, greater than,* and *greater than or equal to*), logical operators (AND, OR, and XOR), conditional and unconditional branching (IF_GOTO and GOTO) and loops (WHILE_DO_ END), apart from the usual mathematical and trigonometric functions. This enables the programmer to write an "intelligent" program that will **automatically** make certain decisions based on certain input conditions. For example, if the specified depth of cut is too large, the machine can be programmed to automatically select a roughing tool. Similarly, appropriate drilling cycle can be automatically selected by the machine depending on the hole parameters. The possibilities are unlimited. Practically any logic that you can think of can be incorporated in a macro program, and then you do not have to worry about a "what-if" type situation. The machine knows what it should do in specific situations.

It is also possible to insert error traps in a program. Under certain specified conditions, the machine will automatically pause or abort the execution of a program and display the programmed error message. After the remedial action is taken by the operator, the paused program can be resumed from that point by pressing the CYCLE START button again. All this makes the program quite flexible, which improves machining quality as well as productivity—an important step toward flexible automation.

Probing

Probing on a CNC machine has several advantages, because the inspection results are immediately available, at a small fraction of the cost of having a separate CMM facility. There is, however, a general perception that since the machine is very accurate, there is no need to measure the obtained dimensions. This is too optimistic a view. The tool may wear out to an unacceptable extent, or it might even break. If the critical dimensions are not checked at least occasionally, a large batch of defective parts might be produced before realizing the problem.

The scope of probing is not limited to just measuring certain dimensions of a part or inspecting the tools for the purpose of quality control. It is also possible to drastically reduce the setup time by automating the tool offset setting procedure. This is extremely useful if the initial sizes of different workpieces are not **exactly** same, which then requires that the offset setting procedure be repeated for every new workpiece.

Probing also obviates the need for extremely accurate fixtures, which are expensive. Any inaccuracy in clamping the workpiece can be measured by the probe, and the corrective logic may be given in the program. In the course of machining, the machine may be programmed to automatically make certain decisions based on probing results. All this leads to better quality and increased productivity at a lower cost. However, probing would not be possible unless the machine has macro programming capability.

Machine Status Information/Manipulation

Many a time it is desirable to know the control conditions of a machine, such as offset distances, the current position of the tool, the current spindle speed/feedrate/tool number, the active modal codes of the various groups of G-codes (e.g., whether G90 or G91 of Group 3 is active on a machining centre. Recall that in every group of G-codes, except those belonging to Group 0 which are nonmodal codes, one code remains active until it is replaced by another code from the same group.) The number of parts produced, the current date/time, the incremental time of machine operation, etc. Such data are made available through a number of *system variables* (described in detail in Chap. 3,) which can be used in a program, for making certain decisions. It is also possible to take the printout of such data on a printer connected to the machine. Some of these variables, such as the variable for storing the current tool position, are read-only types that cannot be modified. Other variables, such as those for offset distances, can be modified through the program, to alter the machine behavior. Such a communication with the machine, to know or alter its control status, is possible only through the macro programming feature.

Communication with External Devices

A *programmable logic controller* (PLC) is an integrated part of the CNC (control) hardware. Generally speaking, the CNC controls the motion of the tool, whereas the PLC controls other machine functions. The PLC can be programmed to control the various automatic systems of the machine, as per our requirements. Most controls use the familiar *ladder diagram* for programming the PLC.

This PLC can also be used to accept input signal or pass on output signal from/to some **external device**, using macro programming features. This allows two-way communication between the machine and the outside world. For example, a sensor can be used to find out whether or not a pallet is loaded (input signal). The machining cycle would not start unless the pallet is loaded. And, if the pallet is not loaded, an external lamp would be flashed (output signal). Finally, when a machining cycle is complete, the part count would be incremented by one and displayed on an external display board. And, when the part count becomes equal to a predefined number, the machine would stop, and an external buzzer would be sounded. Such a sequence of operation would be useful in case of an unattended, automatic loading/unloading on the machine. Similarly, there are several other useful and innovative applications of automation in the production process, which would result in higher productivity and better quality.

The macro programming feature virtually puts "life" in a CNC machine and you can make it respond in the manner you want, in some specific situation. What the machine can be made to do is limited only by the ingenuity of the programmer. Practically, any logic, that one can think of, can be implemented on it. Another issue is that not many people are aware of such features of modern CNC machines. Today, using a CNC machine only for conventional machining applications is like using a 64-bit processor, 2-GB RAM computer only for word processing!

1.3   Does My Machine Have Macro Capability?

This is the most pertinent question at this stage, because unless the machine on which one works has macro capability, there is no point in learning macro programming. Although macro programming has now become a standard feature on most controls, it is still an optional feature on some. So, if a new machine is to be purchased, this feature must be mentioned in the specifications. On an existing machine, execute some macro statement to test whether the machine has this option enabled, in the following manner:

- Set proper conditions for machining (hydraulic pump ON, feed drives enabled, spindle rotation enabled, tailstock extended, chuck closed, and door closed).

- Select *manual data input* (MDI) mode, using the selector switch on the *machine operator's panel* (MOP).

- Press the PROG function key on the MDI panel. If a blank program screen does not appear, press PROG again. The blank screen displays.

```
O0000
%
```

- Type #1 = 1 (or any other "harmless" macro statement, so as not to cause any unexpected machine behavior), followed by the *end of block* (EOB) character (which is a semicolon), and then press the INSERT key, on the MDI panel.
- Press the CYCLE START button on the MOP.

This will assign a value of 1 to the macro variable number 1. If no error message comes, the machine has macro programming capability with full functionality (the control manufacturers do not enable this feature with partial functionality; it will either be available or not available). To be doubly sure, check the value of the defined macro variable on the *offset/setting screen*:

- Press the OFS/SET function key on the MDI panel.
- Press the right extension key (▶) below the display screen, if the MACRO soft key does not appear. Now press the MACRO soft key.

This will show, on a *roll-over* screen (i.e., the first line reappears after the last line, in forward scrolling), the values stored in all the macro variables (001 to 033, 100 to 199, and 500 to 999, typically), which can be viewed using the up/down arrow keys and the page up/page down keys on the MDI panel. The value stored in variable number 001 (which is the same as variable number 1) would be displayed as 00001.000 (values are stored using eight digits), with blank fields for the remaining variables (a blank field indicates an undefined variable), except perhaps those in the 500 to 999 range, which retain the previously stored values even after power OFF. These variables, which are called *permanent common variables*, can, however, be re-defined. Pressing the RESET key will clear all the variables except the permanent common variables. A parameter setting, however, can retain the values of all the variables, even after RESET operation. Different types of variables are described in detail in Chap. 3.

If the machine does not have the macro feature, it will not recognize the macro statement, and will give an error message. In such a case, the company whose control is installed on the machine (such as Fanuc, Siemens, Heidenhain, etc.) has to be contacted. They will enable this option on the machine for a charge. It is interesting to note that the electronic hardware of a particular control version is the **same for all the machines** on which the standard features always remain available. But the control manufacturers enable (through software coding) only those optional features for which they receive additional payment. So, do not worry if your machine is not macro-enabled, just be willing to pay for it. However, nonstandard controls usually do not have macro capability.

1.4 Aim of the Present Text

The macro programming feature is so powerful that its applicability is limited only by your imagination. It has given a whole new dimension to conventional part programming. Unfortunately, the resource material for learning macro programming is not readily available. Moreover, information is available either in handbook forms or as application examples. Handbooks are mainly useful for reference purpose only, to have more details about a certain feature, and the application examples assume basic knowledge of the language. The easiest method to learn would be to study the features of the language one by one, in order of increasing complexity, with a number of suitable examples. Such an approach enables one to understand the concepts simply by self-study. The present text is carefully designed with this approach only, and aims at explaining all the necessary tools and techniques for developing macro programs for common applications. **The present text, however, assumes a basic knowledge of conventional part programming.** Macro programming is only its extension. So, it is necessary to brush up the basic programming concepts before venturing into the fascinating world of macro programming.

The subsequent chapters describe the features of macro programming in detail. Although the general discussion is not brand specific, the specific descriptions and programming examples follow Fanuc's *Custom Macro B* language. The earlier version *Custom Macro A* is outdated and no longer used. Our discussion will generally revolve around Fanuc 0i series controls.

Today, Fanuc enjoys over 50 percent market share in CNC control, worldwide. In India, its share is over 75 percent. This has made Fanuc control the de facto standard, which is the main reason for focusing on Custom Macro B in this text. In fact, many smaller companies have specially designed their controls so as to match the features of Fanuc control. Such controls are referred to as *emulated* Fanuc controls, a number of which are available in the market today. Such controls, however, do not ensure 100 percent similarity with Fanuc control. They still have a market because their prices are much lower compared to original Fanuc control. The emulated Fanuc controls generally do not have macro programming capability.

1.5 How to Use This Book

This book is written in textbook style. It is designed to be read sequentially, from the beginning till the end. Proceed further without skipping any chapter or section (unless suggested otherwise), because this may cause difficulty in understanding certain concepts discussed subsequently. Macro programming is a typical high-level programming language that has to be learnt step by step and thoroughly. The more slowly you move, the faster you will reach your destination, because a superficial knowledge of any programming language may land you in trouble anytime, and you may not even be able to figure out your mistakes.

A simple reading, however, will not be sufficient, because it is not possible to remember everything unless it is practiced rigorously. So, as far as possible, the readers should verify all the programming statements on their machines. For example, if it is stated that #0.5 is equivalent to #1, it can be tested by executing a one-line program, #0.5 = 1; which should assign a value of 1 to variable #1.

Simple statements can be verified in MDI mode also, without creating a new program. The default parameter setting for MDI mode, however, automatically erases the typed program after its execution is over (and it is not possible to save a program created in MDI mode). If some change in the previously typed program is desired, the whole program would need to be typed again. A change in parameter setting would solve this problem; this is explained in Chap. 2. With appropriate parameter settings, MDI program can be deleted or modified only intentionally, through the editing keys on the MDI panel.

Finally, it is important to note that there is no universally accepted standard for macro programming. Although all the reputed control manufacturers offer similar features, there are differences in their programming methodology. This precludes the possibility of portability of macro programs among different controls. In fact, even among the different control versions of the same manufacturer, 100 percent portability would generally not be possible, mainly because of differences in control parameters and system variables in different control versions. What is important is to understand the basic concepts of macro programming technique. The necessary fine-tuning to suit a particular control version can always be done.

Macro programming techniques are not very common knowledge among CNC users, many of whom are not even aware that their machine possibly does have such a programming feature. Unfortunately, macro programming is hyped to be an extremely complex and sophisticated way of part programming. This has caused an undue hindrance in the learning process. The fact is that **macro programming is much simpler than a high-level programming language such as PASCAL**. All that's needed is the willingness to learn and a methodical approach. The author believes that a workable knowledge can be acquired in just one week. Read on and find out for yourself!

CHAPTER **2**

Variables and Expressions

2.1 Macro Variables

A *macro variable* is a mathematical quantity that can assume any value within its allowed range. Although macro programming on Fanuc and similar controls has several features of a high-level computer programming language, it is somewhat in a primitive stage when it comes to the way variables are defined. It does not allow the use of arbitrary combinations of alphanumeric characters for designating a variable, for which there is a single specific way. Variables are designated with the # symbol, followed by a number (called the variable number), in the permissible range (which depends on the control version). Some examples of variables are

```
#1
#10
#100
#1000
#10000
```

These numbers represent specific memory locations that may contain some positive or negative arithmetic values (if defined, i.e., if some value is assigned to them) or be empty (if not defined).

2.2 Macro Expressions

There are two types of macro expressions: *arithmetic expressions* and *conditional expressions*.

An arithmetic expression is a mathematical formula involving *variables* and/or *constants* (such as 0.12, 1.2, 12, 120, etc.), with or without *functions* (such as SIN, ACOS, SQRT, LN, etc., which are described in Chap. 4). A *nested* expression and the *argument* of a function must be enclosed within **square brackets**. Small brackets (parentheses) cannot be used because they are used for inserting *comments* in a program. An arithmetic expression evaluates to an arithmetic value.

Examples:
```
1 + #2
#3 + #4 * SIN[30 * [#5 / 10]]
```

A conditional expression includes *conditional operators* (such as EQ, NE, and LT) between two variables/constants/arithmetic expressions. It must be enclosed in square brackets. It evaluates to either TRUE or FALSE (these are called Boolean values). A conditional expression is also referred to as a *Boolean expression*.

Examples:
```
[#1 EQ 0]
[[#2 + 1] LT [SIN[30] * #3]]
```

The available macro variables cannot be used to store the result of a conditional expression. Only arithmetic values can be stored in macro variables. **The Boolean TRUE/FALSE is not equivalent to the arithmetic 1/0.**

Examples:
```
#1 = 10;        (Stores 10.000 in variable #1)
#2 = [1 LT 2];  (Illegal, as TRUE or FALSE cannot be stored in a variable)
```

The manual data input (MDI) panel usually has just one key for the left bracket and another one for the right bracket. While typing, whether the bracket will appear as a square bracket or a parenthesis, depends on your parameter setting. If the machine has Fanuc 0i or 0i Mate control, set the first bit (from right), which is conventionally referred to as bit #0 (the eight bits are designated as #0 through #7, starting from the right), of parameter number 3204 to 0 for a square bracket, and 1 for a parenthesis:

	#7	#6	#5	#4	#3	#2	#1	#0
3204								

The default setting for the remaining bits is 0, which is not shown, for the purpose of highlighting what requires to be edited (i.e., bit #0). Obviously, it would not be possible to use square brackets and insert comments **at the same time** with this parameter setting. If some comments are desired to be inserted in a macro program, first type/edit the program using square brackets, wherever required. Then change the parameter setting to get parentheses, for the purpose of inserting comments. After inserting comments, if again some corrections in the program involving square brackets are needed, another change in parameter 3204 would be required.

There is, however, a way to use both types of brackets, with the same parameter setting. For this, set parameter 3204#2 to 1 (3204#0 should remain 0; status of the other six bits does not affect this feature):

	#7	#6	#5	#4	#3	#2	#1	#0
3204						1		0

Such a parameter setting displays an extended character set, as soft keys, in the EDIT mode, displaying "(", ")" and "@". With this setting, if square brackets are needed, use the bracket keys on the MDI panel, and if parentheses are needed, do the following (in EDIT mode):

- Press PROG on MDI panel (press PROG again if the current program is not displayed).
- Press the OPRT soft key.
- Press the right extension key (▶) twice.
- Press the C-EXT soft key.

After this, soft keys for left parenthesis, right parenthesis, and @ (which can be used in the comments inserted in a program) will appear which can be used as and when required for editing in EDIT mode. However, a change in display screen will make these soft keys disappear. If they are needed again, the process to display them will have to be repeated.

Note that the MDI panel and the LCD screen (color or monochrome) come as an integral unit as well as separate units. The stand-alone type MDI panel is larger and has more keys. So, there is a lesser need to use the SHIFT key for typing alphanumeric characters/arithmetic operators, which makes typing faster. The keypad also has separate keys for both types of brackets, obviating the need for displaying soft keys for them.

Coming back to the discussion about macro expressions, an arithmetic expression evaluates to a positive or negative arithmetic value, following the usual priority rule: bracket (innermost first) → function evaluation → division and multiplication → addition and subtraction.

Mixed mode arithmetic, that is, calculations involving both real and integer numbers, is permitted, which results in a real number. In fact, the control stores even an integer number as a real number, with zeroes after the decimal point. So, for example, 10 is equivalent to 10.0, and these can be used interchangeably in all arithmetic calculations. Note that this statement is valid only for arithmetic calculations. For example, X10 may be interpreted as 10 mm or 10 μm (in millimeter mode), depending on whether parameter 3401#0 is 1 or 0.

The negative of the value stored in a variable can be used by putting a minus sign before the variable address, but two **consecutive** arithmetic operators (+, −, *, and /) are not allowed. In such cases, square brackets will have to be used:

```
- #1 + 10       (A valid expression)
10 + -#1        (An invalid expression)
10 + [-#1]      (A valid expression)
```

A variable can also be designated in terms of another variable or an arithmetic expression, which must be enclosed within square brackets:

```
#[#1]           (#1 should contain a number in the permissible range. If,
                for example, #1 contains 10, then #[#1] is equivalent to #10.
                ##1 is illegal)
#[#1+10]        (If #1 contains 10, then the referred variable is #20)
```

The designation number of a variable is an integer quantity. If a real value is specified (which is only a theoretical possibility, as this situation would never arise in any practical application), it is automatically rounded to the nearest integer (refer to Sec. 2.3 for the methods of assigning a value to a variable):

```
#1 = 1.4999999;     (Stores 1.4999999 in variable #1)
#[#1] = 1.5000000;  (Stores 1.5000000 in variable #1)
#[#1] = 1;          (Stores 00001.000 in variable #2. A variable
                    stores and displays a value using eight digits,
                    with three digits after the decimal, unless the
                    value cannot be expressed in this format)
#0.4999999 = 1;     (Illegal command because it tries to assign a
                    value to variable #0. Variable #0 is a pre-
                    defined, read-only type null variable, which
                    cannot be redefined. Properties of null vari-
                    ables are described in detail in Sec. 2.5)
#0.5000000 = 1;     (Stores 00001.000 in variable #1)
```

2.3   Assigning a Value to a Variable

A value can be assigned to a variable in the general format

```
#i = <some value or arithmetic expression>;
```

where i is the variable number. On the left-hand side, in place of i, an expression may also be used. Some examples are

```
#1 = 10;            (Stores 00010.000 in variable #1)
#1 = #1 + 1;        (Redefines #1 by storing 00011.000 in it)
#[#1 + 10] = [10 + 20]/ 30 - #1;
                    (Stores –00010.000 in variable #21)
#[#1] = SQRT[-#21]; (Stores 3.1622777, the **rounded value up
                    to eight digits**, in variable #11. Rounding
                    is **automatically** done by the control, in
```

```
#[#1] = SQRT[-#21];
```
all cases of assigning a value to a variable. Only eight decimal digits are stored)

```
G00 W#11;
```
(Causes a displacement of 3.162 mm, the **rounded value up to three digits after decimal**, as the least input increment for displacement is 0.001 mm, in millimeter mode. In an *NC statement*, rounding of axis values, up to the least input increment of the machine, is automatically done by the control, **if the values are specified in terms of variables**. Refer to Sec. 4.4 also)

The term "NC statement" has been used without formally defining it. It is a program *block* involving at least one NC *address*, such as G, M, F, S, T, X, Y, and Z, except codes for calling a macro program (such as G65, G66, etc.) On the other hand, a *macro statement* simply assigns a value to a variable (#i = <some value or an arithmetic expression>), or jumps to a specified block number (GOTO_ and IF_GOTO_), or uses a conditional statement (IF_THEN_, WHILE_DO_, and END_), or calls a macro program. To put it simply, a macro statement does not directly cause physical machine movement, whereas an NC statement directly controls the machine movement. An NC statement may or may not use macro variables/functions.

There are two major differences in the way the control treats NC statements and macro statements:

- If the program is executed in the single-block mode (there is a switch for this purpose on the MOP), its execution stops at the end of each NC statement, and proceeds to the next block only after the CYCLE START button is pressed again. However, the execution does not stop at the end of a macro statement, it proceeds to the next block. If it is desired to execute the macro statements also in single-block mode, set parameter 6000#5 to 1. In a normal situation, such a requirement would never arise, because a macro statement does not involve machine movement. However, in case of an error in the program, execute the macro statements one at a time to check the intermediate calculations.

- Although the program execution is block by block, the control prereads the next block and interprets it in advance, to speed up the execution. In the radius compensation mode, two blocks are preread, because the control needs to position the tool properly at the end of the current block, to suit the next path segment. However, **all** (or as many as possible) sequential macro statements are read and evaluated immediately. In fact, the control does not count a macro statement as a block. An NC statement constitutes a block.

Coming back to the discussion on defining variables, note that a variable always stores a value with minimum **three decimal places**, if the total number of digits does not exceed **eight**. If less than three decimal places are used, zeroes are added.

Examples:
```
#1 = 1234;          (Stores 01234.000)
#1 = 12345;         (Stores 12345.000)
#1 = 123456;        (Stores 123456.00)
#1 = 1234567;       (Stores 1234567.0)
#1 = 12345678;      (Stores 12345678)
```

If more than eight digits are specified, the additional digits **might** be converted to 0, after rounding up to eight digits (irrespective of decimal position), which may give **unexpected** results, as explained in the following example:

```
#1 = 123456.789;    (Stores 123456.790 in variable #1)
#2 = 123456.794;    (Stores 123456.790 in variable #2)
#3 = #2 - #1;       (Stores 0.000 in variable #3)
```

However, in the Fanuc 0i control, specifying more than eight digits, **for any value**, generates an error message, "TOO MANY DIGITS," and terminates the program execution. It will not store values like 123456.789 in a variable, and will display the error message. If, however, more than eight digits result after an arithmetic calculation, rounding is automatically done up to eight digits, and no error (alarm message) or warning (operator's message) appear.

A variable can also be defined in a conditional manner (conditional statements are explained in Sec. 5.3):

```
#10 = 10;
#25 = 5;
IF [#10 GT #25] THEN #25 = #25 + 10;    (TRUE condition, so
                                         #25 becomes 15.000)
```

2.4 Display of Variables

It is necessary to understand how the values of variables are displayed on the macro variable screen, because it might cause some confusion. The calculated value of a variable must lie within the permitted range ($10^{-29} \leq$ magnitude $\leq 10^{47}$, or be 0). However, all the legal values cannot be displayed correctly on the screen, which uses a simple eight-digit decimal format, without exponential digits. Even then, the value (provided it is legal) held in the variable is **correctly used for further calculations**. Also, Fanuc 0i series controls do not allow more than eight digits (including leading or trailing zeroes) for specifying a value in direct assignment. Some examples are given below:

```
#1 = 0.00000001;              (Illegal value, as it contains
                               more than eight digits)
```

`#1 = .00000001;`	(Assigns **correct value** to #1, but displays nine stars, *********, because the number cannot be correctly displayed using eight digits, because the display automatically adds at least one zero to the left of decimal. So, it should display 0.00000001, which it cannot because nine digits are required. Note that the value is the **same** as the value in the previous example)
`#1 = #1 *10;`	(Assigns 0.0000001 to #1 which is also correctly displayed)
`#1 = #1 * #1 * #1 * #1 / 10;`	(Assigns 10^{-29} to #1, but displays nine stars)
`#1 = #1 / 10;`	(The calculated value is less than 10^{-29}, which does not lie in the permitted range, so the execution terminates with an alarm. Note that the alarm message would be "CALCULATED DATA OVERFLOW," though it is actually a *mathematical underflow*)
`#1 = 10000000;`	(Displays 10000000)
`#1 = #1 * #1 * #1 *#1 * #1 * #1;`	(Assigns 10^{42} to #1, but displays nine stars)
`#1 = #1 * 100000;`	(Assigns 10^{47} to #1, but displays nine stars)
`#1 = #1 * 10;`	(The calculated value is more than 10^{47} which does not lie in the permitted range. So, the execution terminates with an alarm, "CALCULATED DATA OVERFLOW")

2.5 Real versus Integer Values

Though perhaps inappropriate, Fanuc control is very liberal in the use of real numbers in place of integer numbers, and vice versa, **in macro statements** (only). While integer numbers in place of real numbers do not cause any problem, real numbers (direct assignment or the value of an arithmetic expression) used in place of integer numbers are automatically rounded. This has already been explained with reference to designation numbers of variables. Here are some different examples (once again, this is only a theoretical discussion, with the sole purpose of explaining the logic being followed by the control, as it might help in error diagnosis):

```
#1 = 1000.4999;
M03 S#1;          (Equivalent to M03 S1000)
```

```
#1 = 1000.5000;
M03 S#1;                (Equivalent to M03 S1001)
M03 S1000.0;            (An illegal statement. S-value must be an integer
                        number)
#1 = 3.4999999;
M#1 S1000;              (Equivalent to M03 S1000)
#1 = 3.5000000;
M#1 S1000;              (Equivalent to M04 S1000)
M3.0 S1000;             (An illegal statement. M-value must be an integer
                        number)
```

Although the last statement is illegal, if 3.0 (or some other real number or expression) is enclosed within square brackets, it becomes a macro expression, and rounding is done:

```
M[3.0] S1000;           (Equivalent to M03 S1000)
M[3.4999999] S1000;     (Equivalent to M03 S1000)
M[3.5] S1000;           (Equivalent to M04 S1000)
```

This also applies to all other addresses (except G-codes) such as spindle speed also:

```
M03 S[1000.0];          (Equivalent to M03 S1000)
M03 S[1000.4999];       (Equivalent to M03 S1000)
M03 S[1000.5000];       (Equivalent to M03 S1001)
```

All these square brackets can also contain arithmetic expressions. Note that if parameter 3451#2 (on a milling machine only; this parameter is not available on a lathe) is set to 1, the spindle speed can have up to one decimal point, though the interpreted speed would be the rounded integer value:

```
S1000.5;     (Equivalent to S1001, if parameter 3451#2 = 1)
S1000.50;    (Illegal, because of more than one decimal place)
```

Too much flexibility makes a programming language rather "undisciplined," and it becomes error-prone. An inadvertent mistake by the programmer might be interpreted "correctly" by the machine, leading to undesirable consequences. And this is very possible because macro programming may involve complex calculations and complicated logic. In the author's opinion, a real value must not be accepted where an integer value is required. Moreover, as has been described in Sec. 2.6, Fanuc control (in fact, perhaps all controls) treats a null variable as a variable having 0 value in arithmetic calculations. This is again illogical and may prove to be dangerous. Typing mistakes are always possible. Perhaps the control manufacturers should modify their macro compilers to make it a PASCAL-like disciplined language. Presently, the programmer has to very meticulously understand the logic followed by the macro compilers.

The problem is more severe in the use of G-codes. The control does not allow any macro expression as the number of a G-code. So, a command such as G[01] is not equivalent to G01, and is illegal. But, one must be aware that the control may not give an error message in all cases. For example, G[03] causes an unexpected tool movement at rapid traverse rate on 0i Mate TC control, which may cause an accident on the machine. So, as a rule of thumb, **never use a macro expression as the number of a G-code**.

G-codes can, however, use macro variables as their numbers. These variables also can be defined in terms of an arithmetic expression. Though no practical application may ever need to use expressions for defining these variables, one should be aware of even theoretical possibilities, which might be helpful in interpreting the outcome of certain mistakes in the program:

`#1 = 0;`	(Stores 0.000 in #1)
`#2 = 1;`	(Stores 1.000 in #2)
`IF [#2 GT #1] THEN #3 = 99;`	(TRUE condition, so #3 becomes 99.000)
`#4 = [#2 + #3] / 2;`	(Calculation makes #4 equal to 50.000)
`G#1;`	(Equivalent to G00)
`G#2;`	(Equivalent to G01)
`G#3;`	(Equivalent to G99)
`G#4;`	(Equivalent to G50)

Except a few (e.g., G12.1 and G13.1), all G-code numbers are integers. So, the values of the variables (directly assigned or assigned through an expression), to be used as G-code numbers, must be integers. Fanuc control accepts real numbers also, as long as **its rounded form, up to the first place after the decimal point, contains 0 after the decimal point.** In other words, it allows an "error" of ±0.05. So, a number in the range 1.95 $\leq n \leq 2.0499999$ would be taken as 2 (as the rounded form, up to the first place after decimal, is 2.0); in 2.95 $\leq n \leq 3.0499999$ range, it would be taken as 3, whereas 2.05 $\leq n \leq 2.9499999$ is an illegal range (as the range is 2.1 $\leq n \leq 2.9$, in the rounded form). So, 1.95, 2.0, 2.0499999, etc. are all taken as 2, while 1.9499999, 2.05, 2.5, 2.9499999, etc. would be illegal:

`#1 = 1.9499999;`	
`#2 = 1.95;`	
`#3 = 2.0;`	
`#4 = 2.0499999;`	
`#5 = 2.05;`	
`G#1;`	(Illegal)
`G#2;`	(Equivalent to G02)
`G#3;`	(Equivalent to G02)
`G#4;`	(Equivalent to G02)
`G#5;`	(Illegal)

The purpose of explaining the effect of specifying a real number, in place of an integer, is to fully describe the logic used by typical controls. Logically, a situation where the result of calculation (or direct assignment) comes out to be a real number (with nonzero digits after the decimal point), where an integer is expected (such as the values associated with G-, M- and S-codes), would never arise unless there is some mistake in the program. However, **for a quick diagnosis, it is necessary to fully understand the logic built in the macro compiler.** Even if the program is perfect, typing mistakes can always creep in.

Recall that parameter 3401#0 decides whether a **dimensional value** (such as distance and feedrate) expressed as a number without a decimal point would be interpreted as a millimeter value or a micron value (assuming G21 mode). This may cause a serious problem on some controls, other than Fanuc, if an integer number is used in place of a real number, while defining a variable. For example,

```
#1 = 10;
G00 W#1;
```

may cause a displacement of 0.01 mm. So, to be on the safer side, always use decimal points for all dimensional values. This problem, however, does not exist on Fanuc control which would store 00010.000 in variable #1, **irrespective of parameter setting**.

Variables can be used as the values following any letter address except O (program number) and N (sequence number)./(optional block skip number) also does not allow use of variables:

```
#1 = 1;
O#1;              (An illegal statement. Use O1)
N#1 G01 X100;     (An illegal statement. Use N1)
/#1 G01 X100;     (An illegal statement. Use /1)
```

2.6   Undefined Variables

If a value is not explicitly assigned to a variable, it remains undefined, and does not contain anything. Such a variable is called a *null variable* or a *vacant variable*. #0 is a predefined, read-only-type null variable. No data can be stored in it. It has been provided for the sole purpose of *logical comparison* of some other variable with it, to find out whether or not the variable is defined. Comparison statements are used in conditional expressions, which are explained in more detail in Secs. 4.1 and 5.2.

Null Variables in Word Addresses

When a null variable appears in an address, that particular address (only) is **ignored**. For example, if #1 is a null variable, then G00 X#1 will not cause any movement, though G00 will become the active modal code of Group 1. This statement is equivalent to a sole G00 command, without any axis word. On the other hand, G00 X#1 Z100 is equivalent to G00 Z100.

Null Variables in Arithmetic Operations

Usually, if a null variable is used somewhere, for a purpose other than a comparison to find out whether or not it is a null variable, an error message, UNDEFINED VARIABLE, would have been more appropriate. But, the Fanuc macro compiler (and possibly compilers of other manufacturers also) is not designed this way. In **arithmetic operations** such as addition and multiplication, as well as in **function evaluation**, a null variable is treated as a variable having 0 value. In the following examples, all variables are assumed to be null variables, initially:

#2 = #1;	(#2 remains a null variable because there is no arithmetic operation or function evaluation in this statement)
#3 = [#1];	(Presence of brackets does make it an expression, but absence of any arithmetic operation with #1 keeps it null. So, #3 remains a null variable)
#4 = #1 + #1;	(Equivalent to #4 = 0 + 0. So, #4 gets defined and stores 0)
#5 = #1 * 5;	(Equivalent to #5 = 0 * 5. So, #5 gets defined and stores 0)
#6 = 1 / #1;	(Equivalent to #6 = 1 / 0, which will give "DIVIDED BY ZERO" error)
#7 = SQRT[#1];	(Equivalent to #7 = SQRT[0]. So, #7 gets defined and stores 0)
#8 = #4109;	(Variable #4109 contains the currently active feedrate on the machine, and if no F-word has been used anywhere in the preceding program blocks, it contains 0. Hence, #8 would store either the current feedrate or 0. #4109-type variables are described in Chap. 3. These variables are **never null**, even if not explicitly defined)

Null Variables in Conditional Expressions

Except for EQ (equal to) and NE (not equal to), a null variable is the same as 0. In the following statements, #1 is a null variable:

[#1 LT 0]	(Equivalent to 0 LT 0, hence FALSE)
[#1 LE 0]	(Equivalent to 0 LE 0, hence TRUE)
[#1 GT 0]	(Equivalent to 0 GT 0, hence FALSE)
[#1 GE 0]	(Equivalent to 0 GE 0, hence TRUE)
[#1 LT #0]	(Equivalent to 0 LT 0, hence FALSE)
[#1 LE #0]	(Equivalent to 0 LE 0, hence TRUE)
[#1 GT #0]	(Equivalent to 0 GT 0, hence FALSE)
[#1 GE #0]	(Equivalent to 0 GE 0, hence TRUE)

However, a null variable is not the same as 0 in comparison statements using EQ and NE:

```
[#1 EQ 0]    (With EQ, a null variable is not the same as 0, hence FALSE)
[#1 NE 0]    (With NE, a null variable is not the same as 0, hence TRUE)
[#1 EQ #0]   (Both are null variables, hence TRUE)
[#1 NE #0]   (Both are null variables, hence FALSE)
```

2.7   A Simple Use of Macro Programming Features

Finally, the following simple example shows how the use of variables makes the same program work under different requirements:

```
#1 = 0;
#10 = 1;
#100 = 50;
#101 = 60;
#102 = 30;
G#10 X#100 Z#101 R#1 F#102;
```

These statements are equivalent to G01 X50.000 Z60.000 R0.000 F30.000, which can be made to do different things simply by changing the values stored in the different variables. There can be several such statements in the program. There is no need to modify any of them. Just define the variables at the beginning of the program, and change them as per requirement. This is the simplest most, yet effective use of macro programming, which does not require an in-depth knowledge of this programming language. One need only be aware of the types of macro variables and their permissible ranges.

2.8   Retaining Programs in MDI Mode

All the statements made above and in the subsequent chapters have actually been verified on Fanuc 0i Mate TC wherever applicable. If one wishes to verify them, or wants to try some other combinations, this can be safely done using the *dynamic graphic* feature in AUTO mode (*automatic operation mode,* which is also called *memory mode*). If, however, no tool movement is involved, verification can be done without using dynamic graphic, in memory mode. In fact, even MDI mode can be used. But the default setting for the MDI mode is such that the program is erased automatically after its execution is complete. This means that the program would have to be typed again, for a subsequent execution. This makes trial and error impractical in MDI mode. However, if parameter 3204#6 is set to 1, the MDI program is retained even after its execution is over. The RESET key will, of course, erase it. If it is desired to retain it even after pressing RESET key, set parameter 3203#7 to 0. On other versions of Fanuc control, these parameter numbers might be different. Reference to respective parameter manuals would be required. An MDI program, however, cannot be permanently saved with a program number for future use. Switching off the machine erases it permanently.

CHAPTER 3

Types of Variables

3.1 Local and Global Variables

In a high-level computer language, there are concepts of *local variables* and *global variables*. Local variables are defined inside a subroutine, and are *local* to them. They are used for intermediate calculations in the subroutine and have **no significance outside it or in another subroutine nested to it**. So, even if different subroutines use local variables with the same names, these refer to **different and independent memory locations**, and hence do not affect one another. And, in the case of nesting, after the execution of the nested subroutine is over and the execution goes back to the calling subroutine, the previously defined local variables of the calling subroutine again become available to it, with the **same previously stored values**. The main program also can have its own set of local variables. Global variables, on the other hand, are *global* in nature in the whole program. They can be used anywhere with the same meaning since these refer to **the same memory locations**:

```
Main program
    GLOVAR = 1
    Call subroutine 1
    Print GLOVAR
```
(Prints 3, because the global variable, GLOVAR, gets modified by both the subroutines)

```
End
Subroutine 1
    Local variables: LOCVAR
    LOCVAR = 1
    GLOVAR = GLOVAR + 1
    Call Subroutine 2
    Print LOCVAR
```
(Prints 1, not 10, because the local variables LOCVAR of the two subroutines are stored at **different and independent** memory locations)

```
    Return
```

```
Subroutine 2
   Local variables: LOCVAR
   LOCVAR = 10                         (LOCVAR defined here is differ-
                                       ent from LOCVAR defined in
                                       Subroutine 1)
   GLOVAR = GLOVAR + 1
Return
```

Subprograms and macros are used as subroutines in a CNC program. **The discussion about local variables applies to macros only.** A
subprogram does not have its own set of local variables. It uses the variables of the calling program, with the same meaning. And, if a
subprogram defines some new variables, these become available to the
calling program also, when the execution returns to it. In fact, **a subprogram can be considered to be a part of the calling program only**, which
is defined separately to avoid multiple typing of repetitive lines.

In macro programming terminology, local variables are referred to
as *local variables* only, but the global variables are referred to as *common
variables*. Further, common variables of a special type are available,
which retain the values stored in them even after the machine is switched
off. The values stored in these variables remain available **in all the
future machining sessions** and can be used/modified by **all the programs**. These can be cleared (made null) or modified only intentionally.
Such variables are called *permanent common variables*. These variables
are one of the unique features of macro programming, and are not available in conventional programming languages such as PASCAL.

There is one more type of variable that is used to read/write a
variety of control data, indicating machine status, such as current tool
position and tool offset values. These are called *system variables*, some
of which are read-only type. System variables are described in detail
in Sec. 3.5.

Thus, macro variables are of the following types:

- Predefined, read-only null variable (#0)
- Local variables (#1 to #33)
- Common variables (#100 to #199)
- Permanent common variables (#500 to #999)
- System variables (#1000 and above)

Variables #34 to #99 and #200 to #499 are not available and cannot
be used.

3.2   Effect of System Reset on Macro Variables

Whenever M02 or M30 is executed, or the RESET button on the MDI
panel is pressed, all local variables (i.e., #1 to #33) and common variables
(i.e., #100 to #199) are cleared to null. This means that a common variable,

with the same meaning, can only be used in **one program** (and in the subprograms/macros nested to it). So, if a calculated value in one program is needed to be used by other program(s) also, it has to be stored in a permanent common variable (i.e., #500 to #999).

A permanent common variable always has the same meaning for all the programs, not only in the current machining session, but also in all future sessions, until it is modified to store a new value. And, it can be modified anytime by any program. It can also be modified in the MDI mode. It is not cleared by the M02/M30/RESET button.

Although not recommended (in the interest of a safe programming practice, as most of the people believe that the local variables and the common variables in a new program start with null values), the following parameter setting would retain the values stored in local and common variables, even after system reset (though power OFF will still clear them):

Parameter 6001#6 = 1   (Retains common variables even after system reset)

Parameter 6001#7 = 1   (Retains local variables even after system reset)

The default setting for these parameter bits is 0, which clears the stored values, if any, whenever the system is reset. It is, however, good programming practice not to depend on defaults and set all the local and common variables to null before using them in a program. And, the permanent common variables also should be carefully used because they may contain some previously stored values.

3.3   Levels of Local Variables

A macro always starts with null values for all the local variables, except those whose values are passed on to the macro through the arguments of the macro call. The method of passing desired values to certain local variables is discussed in detail in Chap. 7. At this stage, it is sufficient to know that G65 P2 A5 is a macro call for program number 2, with macro variable #1 = 5 (initially), and the other local variables of the macro remaining null.

The local variables, defined in a macro, remain accordingly defined only in the **current** call of the macro. Any subsequent call of the same macro will again start with null values for all the local variables (except, of course, those defined through the argument list):

Main program

00001;   (Program number 1)

G65 P2 A5;   (The macro execution starts with #1 = 5, and sets #10 = 5. Then, it redefines #1 and stores 6 in it. Other local variables of the macro remain null. Once the execution of the macro is complete, the values assigned to #1 and #10 are lost)

G65 P2 A5; (This macro call **does not get affected by the previous call** in any manner. The execution of the macro starts with #1 = 5 and null values for the remaining local variables, as in the previous call. So, the macro sets #10 = 5 and #1 = 6, as in the previous call)

M30; (End of the main program)

Macro

O0002; (Program number 2)

#10 = #10 + #1; (Recall that a null variable is equivalent to 0, in arithmetic operations)

#1 = #1 + 1; (Redefines #1)

M99; (Return to the calling program)

Nesting of macros up to a maximum of four levels is allowed. This means that the main program may call macro 1, macro 1 may call macro 2, macro 2 may call macro 3, and macro 3 may call macro 4. If a local variable with the same designation is being used everywhere, it will have **five different meanings**, corresponding to the main program and the four macros. When a macro calls another macro, a new set of local variables becomes active. However, the values stored in the local variables of the calling macro are not lost. After M99, when the control returns to the calling macro, the previously stored values become available. For example, let us assume that the main program and the four nested macros set a local variable #10 equal to 1, 2, 3, 4, and 5, respectively. Then, when the control returns to, say, macro 2 (after executing M99 of macro 3), the value of #10 would be restored to 3. Note that the values of the local variables of a macro are retained **only in case of nesting**, not in a subsequent call of the macro, as already discussed.

The five different sets of the local variables in nested macros are referred to as five *levels* (which are actually five different memory locations) of local variables. The level of the main program is defined as level 0, and the levels of the subsequent nested macros are incremented by 1:

Main program: #1 to #33 of level 0
Macro 1 (called by main program): #1 to #33 of level 1
Macro 2 (called by macro 1): #1 to #33 of level 2
Macro 3 (called by macro 2): #1 to #33 of level 3
Macro 4 (called by macro 3): #1 to #33 of level 4

Note that the main program can call both macros and subprograms, and, macros and subprograms can also call each other. Since subprograms also allow a nesting of up to four levels, we can have a mixed nesting of up to eight levels—four for macros and four for subprograms—**in any order**: main program calling macro 1, macro 1 calling macro 2, macro 2 calling macro 3, macro 3 calling macro 4, macro 4 calling subprogram 1, subprogram 1 calling subprogram 2, subprogram 2 calling subprogram 3, and subprogram 3 calling subprogram 4. Several other combinations of the four macros and the four subprograms are also possible, such as main program calling macro 1, macro 1 calling subprogram 1, subprogram 1 calling macro 2, macro 2 calling subprogram 2, subprogram 2 calling macro 3, macro 3 calling subprogram 3, subprogram 3 calling macro 4, and macro 4 calling subprogram 4. There is no restriction on the total number of macros or subprograms in a program; only the maximum level of nesting has to be four or less, **separately** for macros and subprograms. In other words, in case of a mixed nesting, **not more than four macros or four subprograms are allowe**d. However, there is no restriction on the total number of nested calls at **different** places in the same program, as long as the maximum permissible level of nesting is not exceeded at each place.

It should, however, be noted that subprograms do not have their **own set** of local variables. A subprogram uses the local variables as defined by the **calling program** (which can be the main program, a macro or another subprogram), and can also modify those variables. For example, let us assume that the main program calls a macro, and the macro calls a subprogram. If the same local variable #10 (say) is being used everywhere, then #10 of the main program and #10 of the macro will have different meanings (corresponding to level 0 and level 1, respectively), but #10 of the macro will be the same as #10 of the subprogram. If the subprogram modifies #10, it will also overwrite #10 of the macro, because both refer to the **same** memory location (corresponding to level 1).

These concepts are further explained through examples of Figs. 3.1 and 3.2, depicting the maximum possible level of nesting. Refer to Chaps. 6 and 7 for the methods of calling subprograms and macros. For these examples, it is sufficient to know the following:

G65 P2 A3;	(Calls program number 2 as a **macro**, and sets for the macro, **#1 = 3**, with the **other local variables remaining null**, initially)
M98 P2;	(Calls program number 2 as a **subprogram**. The calling program and program number 2, both use the **same set of local variables**)
M99;	(Returns execution to the calling program)

Note that a subprogram is no different from a macro, **structure-wise**. Both (as well as the main program) can use all types of macro variables and macro functions. If a program is called by G65, it becomes a macro, and the rules regarding local variables are followed. The same program, if called by M98, becomes a subprogram. However, usually a program is designed to be used either as a macro or as a subprogram (or as the main program). If it is a macro, then certain local variables will be assigned some values through the argument list in the macro call statement; other local variables will remain null. It is also possible not to assign any initial value to any local variable,

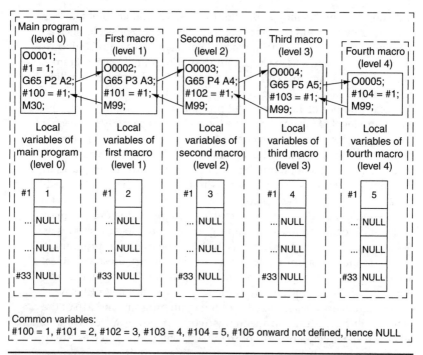

FIGURE 3.1   Local variables in nested macros.

while calling a macro (e.g., G65 P2 which simply calls program number 2, as a macro, with null values for **all** the local variables). On the other hand, a subprogram call is just like a *copy-and-paste* operation of the called program into the calling program. It does not change the level of the local variables of the calling program. So, it can modify the stored value in any variable of the calling program and can also define new variables, during the course of its execution. At the end of its execution, the final values stored in all the variables are passed on to the calling program.

The control stores **all** the programs in the same manner. It is the specific use of a program that classifies it as the main program or a subprogram or a macro. Figure 3.3 gives an example of mixed nesting of a macro and a subprogram. A maximum of three more macros and three more subprograms can be nested to this program. In this program, the main program and the subprogram both refer to the same set of local variables of level 0, whereas the local variables in the macro are level-1 variables. Recall that an undefined (i.e., unassigned) variable is a null variable, which is equivalent to 0 in all arithmetic operations.

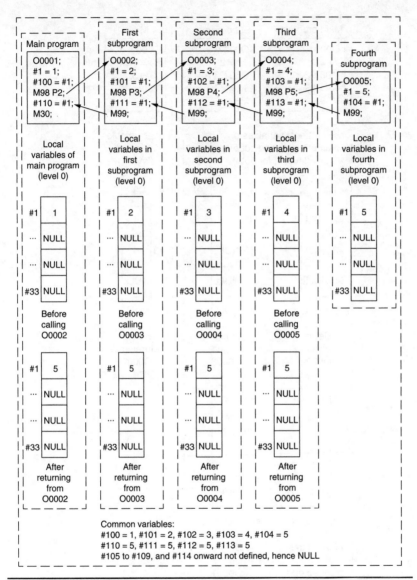

FIGURE 3.2   Local variables in nested subprograms.

3.4   Range of Values Stored in Variables

Local and common variables can have a value 0 or a value in the following ranges:

$$-10^{47} \text{ to } -10^{-29}$$
$$+10^{-29} \text{ to } +10^{47}$$

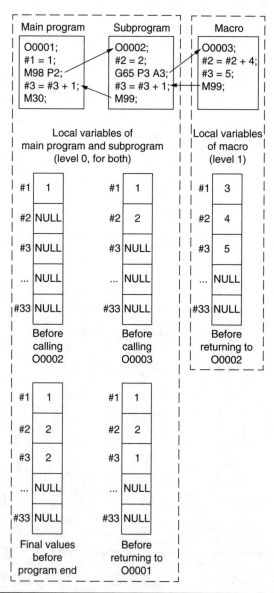

Figure 3.3 Local variable in mixed nesting of macros and subprograms.

If, as a result of calculations, an arithmetic value (final or interme-diate) goes beyond this range, an error condition (which is referred to as an *alarm*) is generated and the execution stops. In fact, a value lying between -10^{-29} and $+10^{-29}$ (but not equal to 0) also is not permit-ted. The associated alarm message is "CALCULATED DATA OVER-FLOW" (though this is actually a case of mathematical underflow).

A programmer may desire that the control should ignore this "error," and treat any value in this range as 0. This, however, is not possible, as there is no parameter to control this feature. But, this is only a theoretical possibility. It is unlikely that such a situation would ever arise in **CNC-related** calculations. An example of the theoretical possibility is given below.

```
#1 = .00000001;
#2 = #1 * #1 * #1 * #1;
```

(Assigns 10^{-8} to variable #1)

(The calculated value is 10^{-32} which generates an alarm, as the value is smaller than 10^{-29})

If such a situation is ever likely to occur, the programmer should normalize the "suspected" variable to 0, whenever its absolute value becomes smaller than, say, 10^{-10}. This will not have any adverse effect on the accuracy of further calculations because macro calculations are accurate up to only about eight decimal digits. A conditional statement such as the following can be used:

```
IF [ABS[#1] LT [0.00001 * 0.00001]] THEN #1 = 0;
```

Note that all the values lying in the permissible range cannot be displayed on the screen, which uses eight-digit decimal format. So, the minimum non-zero value (magnitude) that can be displayed is 0.0000001, and the maximum value is 99999999. The control displays nine stars (*********) as the value of a variable (on the macro variable screen) whenever its display is not possible, provided the value lies in the permissible range (magnitude lying between 10^{-29} and 10^{47}).

3.5 System Variables

The term "system" refers to "CNC control." So, system variables can also be called *control variables*. This is the last group of variables that are related to the **current status** of the CNC control.

System Variables versus System Parameters

System variables should not be confused with *system parameters* (which are commonly referred to as just *parameters*). Parameters decide the **default settings** of the control. For example, as already discussed, the value of a certain parameter decides whether the bracket key on the MDI panel will give a parenthesis or a square bracket. **A parameter is not a variable.** We select a value for parameter to suit our requirement, and do not change it unless our requirement changes. On a new machine, all the parameters are factory-set, keeping in mind the most common applications. Normally, users do not need to change parameters. In fact, they should not, unless it is absolutely essential and the user clearly understands the effect of the change because an incorrect parameter setting may cause unexpected machine behavior, leading to accidents. The users should, in fact, keep the back-up of the original parameters in a safe place so that they could be reinstalled if ever needed in future. Fanuc 0i series controls come with a flash memory card and a PCMCIA slot for it, for backup purposes.

System variables, on the other hand, actually vary, as one works on the machine. Since the values stored in them keep changing, they are called variables, but they cannot be used the way local and common variables are used. The sole purpose of system variables is **to exchange information with the control regarding its current status**. Thus, system variables allow us to communicate with the control, which is essential for general-purpose program development and automation. Such applications, which are the main purpose of macro programming, are discussed in more detail in subsequent chapters.

The only similarity between system variables and system parameters is that both are four-digit numbers (some of these are five-digit numbers also).

System Variables on Fanuc 0i Series Controls

There are hundreds of system variables, and it is unlikely that a user will ever use all of them. He should, however, be well aware of the available information through these variables, without which he cannot fully utilize the capability of his machine. Like system parameters, several of the system variables differ on different control versions of the same company, even for the same functions. The remaining part of this chapter describes some of the commonly used system variables on Fanuc 0i series controls. On a different control, one would need to refer to its *Operator's Manual*. The basic things would be same, only the variable numbers might differ.

Displaying System Variables

Although system variables can be read and written (though some are read-only type) in a program, they are **not displayed on the macro variable screen**. Only the values stored in local and common variables can be seen on the screen. So, if one really wants to see the value stored in a particular system variable, an indirect approach will have to be used: copy the system variable into a local or a common variable, which can then be seen on the offset/setting (OFS/SET) screen.

The information given below is to be used as reference only. There is no need to read everything here, at this stage. So, **skip to the next chapter, after giving a cursory look at this information**. System variables can be broadly categorized on the basis of their use for the following purposes:

- Interface signals
- Geometry and wear offset values
- Workpiece coordinate system shift amount
- Macro alarm (execution stop)

- Time information
- Automatic operation control
- Execution pause (which can be restarted with CYCLE START)
- Mirror image information
- Number of machined parts
- Modal information
- Current tool position
- Work offset values

Interface Signals

This is perhaps the toughest concept in macro programming, because it also requires a knowledge of PLC programming. It is only for advanced users. So, the readers may skip it until they have learned the basic features of macro programming.

Interface signals are used for communicating with (i.e., receiving input and sending output, as binary signals, from/to) **external devices**, connected to the CNC machine. This issue is discussed in more detail in Chap. 12. Here, only a broad overview is given, without going into the finer details, so, certain things may not be very clear until the reader refers to the chapter exclusively devoted to this topic. The main objective here is to explain what interface signals are, not how they are used, though some idea of that is also given.

A discussion on interface signals and their use requires an understanding of the hardware architecture of the control (refer also to *Communication with external devices* in Sec. 1.2 where some information is given). The CNC (control) is mainly concerned with toolpath control. The overall control of the machine is through the logic incorporated in a PLC. For example, if the door is open, machining should not start. If, for some reason, machining is required to be done with the door open, the PLC logic will have to be altered, through a change in its ladder diagram. (Ladder diagram or ladder language is one of the methods of programming a PLC. It is assumed that the reader possesses its basic knowledge. If not, one may refer to some book on PLC programming, such as the one by John R. Hackworth, published by Pearson Education.)

Though a PLC is an integral part of the control hardware supplied by the control manufacturers, it is programmed by the machine tool builders to suit particular machine tools. Fanuc calls its PLC *programmable machine control* (PMC). Essentially, both PLC and PMC refer to the same thing.

When a CNC machine is connected to external devices, two-way communication between the CNC, the PMC, and the external devices is needed. The CNC, however, does not communicate with the external devices **directly**. The communication between the two is through the PMC, as shown in Fig. 3.4.

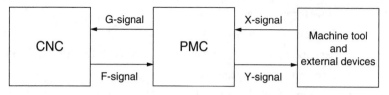

FIGURE 3.4 Communication between CNC, PMC, and external devices.

The PMC accepts inputs from the CNC as well as from external devices (including the machine tool). These inputs are called F-signals and X-signals, respectively. Similarly, it sends outputs to the CNC (G-signals) as well as to the external devices (Y-signals). The X- and Y-signals are also called DI (*data in*) and DO (*data out*) signals, respectively.

Since F- and G-signals are **internal** to the control, the signal addressees and the corresponding meanings are decided by Fanuc. On the other hand, X- and Y-signals are **external** signals, so the signal addresses and their meanings are decided by the MTB. The MTB does not need to use all the available X- and Y-addresses. The left-over addresses can be used by the end-user, as interface signals to/from external devices. Some X-addresses (such as X4.0 to X4.7, and X8.4), however, are standard, and always used with the same meaning.

All the signals are eight-bit signals, each bit carrying a different meaning from the other seven bits. For example, the PMC sends the "cycle start" signal to the CNC by G7.2 (which means bit #2 of G7 or G007; usually, the F- and G-signal addresses are three-digit numbers, and X- and Y-addresses are two-digit numbers, where the leading zeroes can be omitted). On the other hand, G7.1 is the "start lock" signal (which disables the CYCLE START button on the MOP), and G7.0 is not defined. Similarly, X0.0 (bit #0 of X0) and X0.1 (bit #1 of X0) are two independent input signals. Sometimes, a combination of certain bits is used for sending appropriate signals. For example, G43.0, G43.1, and G43.2 are for mode selection signal to the CNC. When these are all 0, the CNC considers it an instruction to work in the MDI mode. On the other hand, 001 is for HANDLE mode, 100 is for MEMORY or AUTO mode, and so on.

The hardware connections for F- and G-signals are internal. The predefined meanings of F- and G-signals cannot be changed by the MTB or the end-user. The address numbers for these signals and the associated meanings can be found in the Fanuc *Maintenance Manual*.

A number of X- and Y-signals are available for interacting with the outside world. Typically, X0 to X3 are for general-purpose signals, X4 to X11 are connected to the keys of the Fanuc-manufactured MOP, X12 is connected to the *manual pulse generator* (MPG), X13 and X14 are for additional MPGs, and X15 is for error signals. Many MTBs do not use Fanuc's MOP. They design their own MOP. In such cases, X0 to X11 become general-purpose signals, and any signal can be used for any purpose. The available Y-signals typically are Y0 to Y7. One will have to refer to the manual supplied by the MTB to find out which X- and Y-addresses have been used by them. The remaining addresses can be used by the users for the purpose of communication with external devices.

The physical connection between the PMC and the external devices is done by connecting wires to the *terminal strip* attached to the I/O (input/output) module (also called the I/O unit) of the PMC. Refer to the *Connection Manual* (*Hardware*) of Fanuc to know the pin assignment (i.e., which terminal corresponds to which X- or Y-address) on the terminal strip. The I/O module is generally placed in the wiring cabinet designed by the MTB. It is connected to the terminal strip by four 50-pin connectors (on the I/O module) and *ribbon cables*.

In order to communicate an external signal to the CNC, the corresponding X-signal will need to be written to the appropriate G-signal (G54.0 to G54.7 and G55.0 to G55.7 are reserved for this purpose). This can be done by **adding a new rung to the PMC ladder**:

This defines a system variable (corresponding to the specified G-address), which can be read in a program. For example, when X0.0 (assuming the MTB has not used this address) is written to G54.0 (which corresponds to variable #1000), it defines variable #1000 (which becomes 0 or 1 depending on whether X0.0 is *low* or *high*). Variable #1000 can be read in a program. Thus, the ON/OFF state of an external sensor becomes available inside the program.

Similarly, the F-signals (those which are used for the purpose of external communication, F54.0 to F54.7 and F55.0 to F55.7) generated by the CNC become available to the outside world only when these are passed on as the Y-signals by adding new rungs to the PMC ladder (then these signals can be tapped from the specified pins on the terminal strip and sent to external devices through wires):

For example, if F54.0 (which corresponds to variable #1100) is written to Y0.0 (assuming the MTB has not used Y0.0), the assigned value (0 or 1) to variable #1100, inside the program, becomes available at the output terminal Y0.0, which can be used to switch an external device on or off. Thus, a program statement can drive an external device.

The binary signals (corresponding to variables #1000, etc.) for such a communication are called interface signals. Typically, 16 input and 16 output signals are available. The current status of these signals is available inside a program through certain system variables, as already explained by some examples.

Two types of system variables are available: for 16 single-bit signals and for one 16-bit signal (for both input and output signals). The examples given pertain to system variables for single-bit signals which can only be 0 or 1. The variables for 16-bit signals are used to read/write all the 16 signals **simultaneously** by a single program statement. For example, if the first and the second signals (i.e., F54.0 and F54.1) are 1 and the rest are 0 then the corresponding 16-bit signal would be 0000000000000011 (the decimal interpretation of which is 3). So, #1100 = 1; and #1101 = 1; is the same as #1132 = 3; except that the two outputs are triggered simultaneously in the second case. It is a matter of individual choice/requirement whether the signals are read/written by the program one by one or simultaneously. The complete variable list is given below. (32-bit signals are also available. They are not described here because they are rarely used.)

System Variables for Input Interface Signals

`#1000 to #1015, #1032:`   #1000 to #1015 are single-bit signals. #1032 is a 16-bit signal, whose individual bits (starting from the right) correspond to #1000 to #1015. These are read-only variables.

System Variables for Output Interface Signals

`#1100 to #1115, #1132:`   #1100 to #1115 are single bit signals. #1132 is a 16-bit signal whose individual bits (starting from the right) correspond to #1100 to #1115. These are read/write variables.

The decimal values of #1032 and #1132 (all other variables are either 0 or 1) would be given by the following equations:

$$\#1032 = \sum_{i=0}^{15} \#[1000+i] \times 2^i$$

$$\#1132 = \sum_{i=0}^{15} \#[1100+i] \times 2^i$$

where the individual terms on the right-hand side would be either 0 or 2^i, depending on whether the corresponding signal is 0 or 1.

The correspondence between these variables and the F- and G-signals is given in Table 3.1.

Geometry and Wear Offset Values

The specified geometry and wear offset values are stored in certain system variables. These variables are described below for recent

G54.0	#1000	G55.0	#1008	F54.0	#1100	F55.0	#1108
G54.1	#1001	G55.1	#1009	F54.1	#1101	F55.1	#1109
G54.2	#1002	G55.2	#1010	F54.2	#1102	F55.2	#1110
G54.3	#1003	G55.3	#1011	F54.3	#1103	F55.3	#1111
G54.4	#1004	G55.4	#1012	F54.4	#1104	F55.4	#1112
G54.5	#1005	G55.5	#1013	F54.5	#1105	F55.5	#1113
G54.6	#1006	G55.6	#1014	F54.6	#1106	F55.6	#1114
G54.7	#1007	G55.7	#1015	F54.7	#1107	F55.7	#1115

TABLE **3.1** Correspondence between Interface Signals and System Variables

versions of 0i series controls that use what is called *Memory Type C.* For an older control version (i.e., for those using *Type A* or *Type B* memory) refer to the machine manual.

Since there is a difference in the tool geometry on lathes and milling machines, the system variables for the various offset distances carry different meanings on these two types of machines. Typically, 64 (#2000 series) or 99 (#10000 series) offset numbers on a lathe (#10000 series can be used for 64 offset numbers also), and 200 (#2000 series) or 400 (#10000 series) offset numbers on a milling machine (#10000 series can be used for 200 offset numbers also) are available. It is, however, better to use the #10000 series, because its range is higher, and it can also be used for the lower range. #2000 series was used in older control versions. #10000 series was introduced for the purpose of increasing the offset numbers, which also allows for further increase beyond 99 (on lathes) and 400 (on milling machines) in future control versions. If an attempt is made to read/write undefined system variables, an error message, "ILLEGAL VARIABLE NUMBER," is displayed. For example, a command such as #1 = #10065; would give an error message on a lathe with 64 available offsets. Tables 3.2, 3.3, and 3.4 show the system variables for various offset values on lathe and milling machines.

When offset setting is done, the corresponding system variables automatically store the respective offset distances. These are read/write variables, so it is also possible to change the offset values by modifying these variables. This can be done by executing, say, #15001 = 5; in MDI or memory mode, which will set the specified value (5, in this example) as the X-axis geometry offset value corresponding to offset number 1, on a lathe.

For modifying offset values during the program execution, the appropriate system variables need not be redefined explicitly. It is possible to modify the offset values (and hence, the associated system variables also) through the *programmable data entry* command, G10,

	Offset Number	Wear Offset Value	Geometry Offset Value
	1	#2001	#2701
	2	#2002	#2702
X-axis offset values	...	...	...
	...	...	...
	64	#2064	#2764
	1	#2101	#2801
	2	#2102	#2802
Z-axis offset values	...	...	...
	...	...	...
	64	#2164	#2864
	1	#2201	#2901
	2	#2202	#2902
Nose radius values	...	...	...
	...	...	...
	64	#2264	#2964
	1	#2301	
	2	#2302	
Tool-tip directions	...	...	
	...	...	
	64	#2364	

TABLE 3.2   System Variables for Lathe Offsets (with 64 Offset Numbers)

which is described in Chap. 13. G10 can also be commanded in the MDI mode.

Since an error in specifying offset values produces bad parts, and a serious error may even cause an accident on the machine, offset modification should not be done during the automatic execution of a program unless its effect is fully verified by first testing the effect of the modification in MDI mode. For this, first execute the offset change command (direct assignment or through G10) and the T-code in MDI mode, and then check the position display after manually bringing the tool to a **known** position. On a milling machine, tool length compensation code (G43/G44) also will have to be executed for verifying the tool length offset value. Tool radius (nose radius and tip direction on a lathe) can simply be seen in the offset tables for verification of its value.

	Offset Number	Wear Offset Value	Geometry Offset Value
	1	#10001	#15001
	2	#10002	#15002
X-axis offset values	...	...	...
	...	...	...
	99	#10099	#15099
	1	#11001	#16001
	2	#11002	#16002
Z-axis offset values	...	...	...
	...	...	...
	99	#11099	#16099
	1	#12001	#17001
	2	#12002	#17002
Nose radius values	...	...	...
	...	...	...
	99	#12099	#17099
	1	#13001	
	2	#13002	
Tool-tip directions	...	...	
	...	...	
	99	#13099	

TABLE 3.3   System Variables for Lathe Offsets (with 99 Offset Numbers)

Workpiece Coordinate System Shift Amount

The workpiece coordinate systems, defined by G54 etc. can be shifted by the desired amounts along the X- and/or Z-axis, through the *workpiece shift screen* on a lathe. Such a facility is **not available on a milling machine**. Press the right extension soft key twice on the offset/setting screen, to get the workpiece shift screen.

Such a shift, which applies to all subsequent operations of the machine until the shift value is changed, is useful for shifting the defined workpiece coordinate system (by the shift amount, in the **opposite** direction) **without changing the offset distances**. For example, if the position display is 100 and –50 (for the X- and Z-axis, respectively, on a lathe), a shift of +1 in both the X- and Z-direction would shift the coordinate system by 1 mm in the negative X- and Z-direction, changing the

	Offset Number	Wear Offset Value	Geometry Offset Value
Tool length offset values (H)	1	#10001 (#2001)	#11001 (#2201)
	2	#10002 (#2002)	#11002 (#2202)
	...	...	...
	200	#10200 (#2200)	#11200 (#2400)
	...	...	...
	400	#10400	#11400
Tool radius values (D)	1	#12001	#13001
	2	#12002	#13002
	...	...	...
	...	...	...
	400	#12400	#13400

Note:
1. This table corresponds to parameter 6000#3 = 0. When this parameter is set to 1, the wear and geometry variable numbers get interchanged.
2. When the available offset numbers are not greater than 200, #2001 through #2200 can also be used in place of #10001 through #10200. Similarly, #2201 through #2400 can be alternatively used for #11001 through #11200.

TABLE **3.4**   System Variables for Milling Machine Offsets (with 200/400 Offset Numbers)

position display to 101 and –49, respectively. The specified shift uniformly applies to **all** the workpiece coordinate systems, defined by G54, G55, etc.

The associated system variables are given in Table 3.5. These are read/write variables. For example, to specify a shift of +1 for the X-axis, set #2501 = 1. Similarly, to read the shift amount for the X-axis, set #1 = #2501, and read the value of variable #1 on the macro variable screen.

Macro Alarms

Alarm refers to an error condition that terminates the execution of the current operation. Pressing the MESSAGE key on the MDI panel

Controlled Axis	System Variable for Shift Amount
X-axis	#2501
Z-axis	#2601

TABLE **3.5**   System Variables for Workpiece Coordinate System Shift Amount

shows the error message. The alarm condition may arise due to a number of reasons, including hardware problems with the machine. Further operation on the machine is possible only after rectifying the problem, and then pressing the RESET key. A syntax error in the program, out of range values, illegal data entry, etc. are software-related error conditions. CNC automatically issues an alarm in such cases.

There are, however, situations when the control will not sense the error, even though the program execution must be terminated. For example, a probing system installed on the machine may detect tool breakage, necessitating immediate termination of program execution, though the control will not recognize this problem. To tackle such cases, provision also has been made to issue user-generated alarms, which are called *macro alarms*.

When a value from 0 to 200 is assigned to system variable #3000, the machine stops with the alarm message, "MACRO ALARM." The displayed alarm number is 3000 plus the number assigned to it. Thus #3000 = 1; would terminate the current operation, and display "3001 MACRO ALARM" when the MESSAGE key is pressed.

It is also possible to display a user-specified message up to 26 characters as alarm message by typing it within parentheses, after the assigned value for variable #3000.

Example:
```
#3000 = 1 (TOOL BROKEN);
```

that would display "3001 TOOL BROKEN" on the message screen.

Time Information

The system variables for time-related information are given in Table 3.6. Variables #3001 and #3002 are read/write variables, whereas #3011 and #3012 are read-only variables. So, it is not possible to change the current date or time through these variables.

Current date/time can only be set/altered on the time setting screen, which is called *timer/part count screen*. (Select MDI mode → Press OFS/SET key → Press SETTING soft key → Press page down key twice → Using up/down/left/right arrow keys, bring the cursor to time or date display, type the new value and press INPUT key.) This screen also displays some time-related information such as power ON time and cutting time, in hour-minute-second format. Part count information (the number of parts required and the number of parts produced) is also displayed on this screen. This is discussed in the section "Number of Machined Parts."

Automatic Operation Control

The single-block switch on the MOP can be used to execute a program block by block, that is, one block at a time. For executing the next block, the CYCLE START button must to be pressed again.

Variable Number	Function
#3001	This variable functions as a timer, with a 1-ms increment. It counts the total "on-time" of the machine in the **current** session. When the power is turned off, the value of this variable is reset to 0. When 2147483648 ms (which is nearly equal to 25 days) is reached, the value of this variable resets to 0, and time counting starts again.
#3002	This variable functions as a timer, with a 1-hour increment. It counts the total "run-time" of the machine (i.e., the duration of on-time of the CYCLE START lamp) in **all** the sessions. When the power is turned off, the value of this variable is preserved. When 9544.371767 hours (which is nearly equal to 13 months) is reached, the value of this variable resets to 0, and time counting starts again.
#3011	This variable stores the current date (year-month-day) in decimal format. For example, 4 February 2009 is represented as 20090204.
#3012	This variable stores the current time (hour-minute-second) in decimal format. For example, 21 minutes and 30 seconds past 8 pm is represented as 202130.

TABLE 3.6   System Variables for Time Information

However, it is possible to disable this switch through the system variable #3003.

Another feature, which is controlled by variable #3003 concerns completion of auxiliary functions (M-, S- and T-code), specified in a block also having a G-code. The G-codes allow specification of auxiliary codes in the same block. Whether the execution of the G-code starts immediately or waits for the completion of the auxiliary command is decided by this variable. For example, if G01 X100 S500; is commanded, the linear interpolation may start immediately, without waiting for the spindle to attain 500 rpm, or it may wait till the specified rpm is reached. Note that the T-code possibly cannot be used for changing the tool during machining because of a possible interference between the tool and the workpiece or the machine body. The only use of the T-code, in such a case, would be for changing the tool offset number (e.g., T0101, T0111, T0112, etc., with tool number 1 being in the cutting position, on a lathe).

Even if the auxiliary codes are specified in a separate block, the execution of the next block may or may not wait for the completion of the auxiliary command, depending on the setting of variable #3003.

#3003	Single Block	Completion of an Auxiliary Function
0	Enabled	To be awaited
1	Disabled	To be awaited
2	Enabled	Not to be awaited
3	Disabled	Not to be awaited

TABLE **3.7** System Variable (#3003) for Automatic Operation Control

Table 3.7 shows the effect of assigning the four permissible values to this variable. When the machine is switched on, the value of this variable is 0, irrespective of its value in the previous machining session.

Another system variable, #3004, controls feed hold and feed override switches on the MOP as well as the exact stop check performed internally by the control at the end of a cutting motion. Pressing the feed hold button stops tool movement without terminating the program execution. The tool movement starts again as soon as the feed hold button is released. The feed override switch allows feedrate from 0 to 254 percent of the programmed value (or less, depending on the particular machine tools), except in threading where it is automatically disabled by the control (i.e., it remains fixed at 100 percent). The exact stop check is performed by the control to ensure that the tool reaches the commanded point within specified tolerance. This slows down the performance of the machine a bit. So, if too much accuracy is not desired for a particular cutting operation (e.g., a rough cutting, which is to be followed by a finish cutting), this check may be temporarily disabled. Table 3.8 shows the effect of variable #3004 on feed hold, feed override, and exact stop check. When the machine is switched on, the value of this variable is automatically made 0 by the control, irrespective of its value in the previous machining session.

#3004	Feed Hold	Feed Override	Exact Stop Check
0	Enabled	Enabled	Performed
1	Disabled	Enabled	Performed
2	Enabled	Disabled	Performed
3	Disabled	Disabled	Performed
4	Enabled	Enabled	Not performed
5	Disabled	Enabled	Not performed
6	Enabled	Disabled	Not performed
7	Disabled	Disabled	Not performed

TABLE **3.8** System Variable (#3004) for Automatic Operation Control

Execution Pause

System variable #3000 generates an alarm condition and **terminates** the program execution, whereas variable #3006 causes temporary **pause** of execution, which can be restarted by pressing the CYCLE START button again. In the paused state, pressing the MESSAGE key displays the user-specified message (up to 26 characters). Assigning a number to variable #3006 halts the program execution. There is no significance to this number, because message number is not displayed. So, normally, 1 is assigned. Example:

```
#3006 = 1 (CHECK THE DIAMETER);
```

This would temporarily stop the execution, and display "CHECK THE DIAMETER" on the message screen. If no message is typed, nothing would be displayed.

Mirror Image Information

Apart from programmable mirror image command, it is also possible to obtain mirror image profiles using external switches on the MOP (there are separate switches for different axes), or through the *mirror image setting screen.* (Select MDI mode → Press OFS/SET function key → Press SETTING soft key → Now, after pressing page down key, the mirror image screen will appear, on which select 1 for activating the mirror image for a particular axis.)

The mirror image status of each axis is stored in a read-only system variable #3007. This is a bit-type variable, but it stores the value in decimal form. For example, 00000011 (where 1 indicates mirror image enabled and 0 indicates mirror image disabled) is stored as 3. So, the value stored in this variable needs to be converted into binary for interpreting it. For example, a value of 3 indicates that mirror image is enabled for the first two axes. Table 3.9 shows the bit settings for this variable.

On a lathe, the first and the second axes refer to the X- and Z-axis, respectively, and on a milling machine, the first three axes refer to the X-, Y-, and Z-axis, respectively. Additional axes may or may not be available on a particular machine. Only the first four bits of this variable are used for the Fanuc 0i series.

Number of Machined Parts

The number of parts required and the number of completed parts in the current machining session can be read or written in system variables

Bit #7	Bit #6	Bit #5	Bit #4	Bit #3	Bit #2	Bit #1	Bit #0
				4th axis	3rd axis	2nd axis	1st axis

TABLE 3.9 Bit Settings for System Variable #3007

#3902 and #3901, respectively. Part count information can also be seen/ modified on the timer/part count screen, or through parameters 6713 and 6711 which store the number of parts required to be produced and the number of parts produced, respectively.

The timer/part count screen also displays the **total** number of parts produced during the entire service time of the machine, as read-only information. Whenever a part is completed, both the number of parts produced and the total number of parts produced are incremented by 1. While the required and the completed number of parts can be altered (through the associated system variables, parameter settings, or the setting screen), as and when desired, the total number of completed parts can only be altered through parameter 6712. This information has been made available to keep a record of the total number of parts produced during the entire service time of the machine.

The default setting (parameter 6700#0 = 0) of the machine increments the part count by 1 whenever M02, M30, or the M-code number specified in parameter 6710 (which normally contains 30 only, signifying M30) is executed. If it is desired to increment the part count only after the M-code specified in parameter 6710 is executed (which may be the same as or different from 02 or 30, but not 0, 98, or 99), set parameter 6700#0 = 1.

Finally, note that M02 or M30 must be typed with the *end-of-block* (EOB) symbol (i.e., as M02; or M30;), otherwise **the part count will not be incremented**, even though the part would be produced without any error message (the control allows the missing EOB symbol at the end of the program).

Modal Information

The G-codes have been categorized into different groups based on similarity in functionality. At any time, one G-code from each group (except Group 0) remains active. Those belonging to Group 0 are called *nonmodal* codes, which remain effective only in the block where they are programmed. Other codes are *modal* codes, which remain effective until replaced by some other code from the same group. For example, G01 is a modal code, grouped with G00, G02, G03, etc. So, if G01 is used once in the program, it need not be typed again, if the subsequent motions are to be executed with linear interpolation. However, as soon as G02 (or some other code from the same group) is commanded, G01 is cancelled and G02 becomes active.

During programming, specially while using general-purpose macros, which are to be used with several programs, it may be necessary to know the active G-code of a particular group. For example, referring to a milling machine, if the main program calls a macro in incremental mode, the macro should interpret the values in the macro call argument list as being incremental values. The macro should work correctly in cases of both absolute and incremental modes of the calling program. This would be possible only if this information is available while executing the macro. The system variable #4003 stores 90 or 91 depending on whether absolute or incremental mode is currently active. The macro can read this to have information about the active mode. For example, let us assume that variable #1 of the macro represents the absolute X-axis position, and it gets a value through the argument list of the macro call. Then, to take care of the possible incremental mode of the main program at the time of calling the macro, the following statement in the beginning of the macro would require to be inserted (IF_THEN_ statement is described in Chap. 5):

```
IF [#4003 EQ 91] THEN #1 = #5041 + #1;
```

The system variable #5041 stores the current position of the tool along the X-axis, as described in Table 3.11.

The system variables for various modal information on Fanuc 0i series are given in Table 3.10 (a) and (b). The vacant columns are for

Variable Number	Function	G-Code Group
#4001	G00, G01, G02, G03, G33, G34, G71–G74	1
#4002	G96, G97	2
#4003		3
#4004	G68, G69	4
#4005	G98, G99	5
#4006	G20, G21	6
#4007	G40, G41, G42	7
#4008	G25, G26	8
#4009	G22, G23	9
#4010	G80–G89	10
#4011		11
#4012	G66, G67	12
#4014	G54–G59	14
#4015		15
#4016	G17, G18, G19	16
#4017		17
#4018		18

TABLE 3.10(a)   System Variables for Modal Information on a Lathe

Variable Number	Function	G-Code Group
#4019		19
#4020		20
#4021		21
#4022		22
#4109	F-code (feedrate)	
#4113	M-code number	
#4114	Sequence number	
#4115	Program number	
#4119	S-code (rpm/constant surface speed)	
#4120	T-code (tool number with offset number)	

TABLE 3.10(a)   (*Continued*)

Variable Number	Function	G-Code Group
#4001	G00, G01, G02, G03, G33	1
#4002	G17, G18, G19	2
#4003	G90, G91	3
#4004		4
#4005	G94, G95	5
#4006	G20, G21	6
#4007	G40, G41, G42	7
#4008	G43, G44, G49	8
#4009	G73, G74, G76, G80–G89	9
#4010	G98, G99	10
#4011	G50, G51	11
#4012	G66, G67	12
#4013	G96, G97	13
#4014	G54–G59	14
#4015	G61–G64	15
#4016	G68, G69	16
#4017		17
#4018		18

TABLE 3.10(b)   System Variables for Modal Information on a Milling Machine

Variable Number	Function	G-Code Group
#4019		19
#4020		20
#4021		21
#4022		22
#4102	B-code number	
#4107	D-code number	
#4109	F-code (feedrate)	
#4111	H-code number	
#4113	M-code number	
#4114	Sequence number	
#4115	Program number	
#4119	S-code (rpm/constant surface speed)	
#4120	T-code (tool number)	
#4130	P-code number of the currently selected additional workpiece coordinate system	

TABLE 3.10(b)   System Variables for Modal Information on a Milling Machine (*Continued*)

other versions of Fanuc. These system variables are, obviously, read-only variables. Note that all the G-codes of a particular control version may not be available on a particular machine. Some G-codes are optional features of the control (which need to be separately purchased), and others cannot be used because of hardware limitation of the machine tool.

The parameter 6006#1 (on Fanuc 0i series milling machine controls) decides whether the modal information obtained through system variables (#4001 to #4022) is up to the immediately **preceding** block (which is the default setting) or up to the **currently executing** block:

```
6006#1 = 0 (up to the immediately preceding block)
6006#1 = 1 (up to the currently executing block)
```

The control automatically highlights the currently executing block on the display screen. This parameter, however, is not available on lathe where model information is available up to the last executed block.

Current Tool Position

The system variables, which store information about the current tool position, are given in Table 3.11. These are all read-only variables. Note the following:

Variable Number	Position Information	Coordinate System	Tool Compensation Value	Read Operation during Tool Movement
#5001–#5004	Block end point	Workpiece coordinate system	Not included	Enabled
#5021–#5024	Current position	Machine coordinate system	Included	Disabled
#5041–#5044	Current position	Workpiece coordinate system	Included	Disabled
#5061–#5064	Skip signal position	Workpiece coordinate system	Included	Enabled
#5081–#5084	Tool offset value			Disabled
#5101–#5104	Deviated servo position			Disabled

TABLE **3.11** System Variables for Current Tool Position

Variable Number	Function	Remark
#5201–#5204	External workpiece coordinate system offset values	
#5221–#5224	G54 workpiece coordinate system offset values	
#5241–#5244	G55 workpiece coordinate system offset values	
#5261–#5264	G56 workpiece coordinate system offset values	
#5281–#5284	G57 workpiece coordinate system offset values	
#5301–#5304	G58 workpiece coordinate system offset values	
#5321–#5324	G59 workpiece coordinate system offset values	
#7001–#7004	1st additional workpiece coordinate system offset values (G54.1 P1)	Optionally available on milling machines
#7021–#7024	2nd additional workpiece coordinate system offset values (G54.1 P2)	
…	…	
#7941–#7944	48th additional workpiece coordinate system offset values (G54.1 P48)	

TABLE **3.12**   System Variables for Work Offset Values

- The first digit from the right (1, 2, 3, or 4) represents an axis number. So, on a 2-axis lathe, 1 represents the X-axis, 2 represents the Z-axis, and the remaining two are not used. This also applies to Tables 3.12 and 3.13.

- The tool position where the skip signal is turned on in a G31 (skip function) block is stored in variables #5061 to #5064. When the skip signal is not turned on during the course of tool movement under G31, the specified end point in this block is stored in these variables.

- Certain variables, such as those for the current tool position, cannot be read while the tool is moving. The values stored in them can only be read when the tool movement stops. So the "current position" is not really the **instantaneous** position.

Function	Variable Numbers	Alternate Variable Numbers
External offsets	#5201–#5204	#2500–#2800
G54 offsets	#5221–#5224	#2501–#2801
G55 offsets	#5241–#5244	#2502–#2802
G56 offsets	#5261–#5264	#2503–#2803
G57 offsets	#5281–#5284	#2504–#2804
G58 offsets	#5301–#5304	#2505–#2805
G59 offsets	#5321–#5324	#2506–#2806

TABLE 3.13 Alternate System Variables for Work Offset Values

Work Offset Values

Refer to Sec. 5.5 for a detailed discussion on how offset distances for various *workpiece coordinate systems* are defined. The measured work offset distances get stored in certain system variables. These are read/write variables. So, it is also possible to change the offset distances by modifying the contents of these variables. Table 3.12 explains the function of these variables. Table 3.13 shows alternate variable numbers that can also be used on milling machines, for work offset distances.

CHAPTER 4

Macro Functions

4.1 Types of Macro Functions

Macro programming is equipped with all the commonly used mathematical functions that are typically available in a high-level computer programming language. The available macro functions can be separated into groups, to make their understanding and usage easier to learn. There are broadly five groups, apart from the usual arithmetic operations:

- Trigonometric functions
- Rounding functions
- Miscellaneous functions
- Logical functions
- Conversion functions

A value may be assigned to a variable (or an expression that evaluates to a legal variable number) using any combination (that should, of course, be logically correct) of these functions. An unassigned (undefined) variable is called a null variable that is equivalent to 0 in mathematical calculations. So, as a general rule, a variable should not be used in any mathematical expression without first defining it.

Priority of Operations in Arithmetic Expressions

In a complex expression, involving several functions and brackets, the following priority rule is followed:

1. Brackets (only square brackets are permitted)
2. Functions (such as SIN, ATAN, SQRT, and EXP)
3. Multiplication (*), division (/), and bitwise AND
4. Addition (+), subtraction (−), bitwise OR, and bitwise XOR

If more than one operations of equal priority appear in a row, the calculation is done from left to right.

Examples:

```
2 + 2 / 2        (Returns 3)
[2 + 2] / 2      (Returns 2)
2 / 2 / 2        (Returns 0.5)
```

Priority of Operations in Boolean Expressions

A Boolean expression can have arithmetic expressions (within square brackets) to the left as well as to the right of the conditional operator. The priority rules are followed inside each bracket independently:

```
[[2 + 2 / 2] LT [2 + 2]]
```
(Equivalent to [3 LT 4], which returns TRUE)

The arithmetic expression at the right of a conditional operator must be enclosed within brackets, but the brackets at the left are optional:

```
[2 + 2 / 2 LT [2 + 2]]
```
(Same as above, hence returns TRUE)

```
[2 + 2 / 2 LT 2 + 2]
```
(After evaluating the arithmetic expression at the left, which returns 3, further evaluation is done from left to right. So, the statement is equivalent to FALSE + 2, which is meaningless and hence illegal)

Skipping brackets in Boolean expressions (even if it does not give an error, in some cases) is a source of confusion. So, it is a good programming practice to use brackets on both sides of conditional operators, as well as outer brackets for the whole expression.

Effect of Order of Calculations

In some cases, the order of calculation does not change the mathematical meaning of an arithmetic expression, and therefore, the same final result is expected. However, in certain cases of extremely large or extremely small values, the order of calculation may become very important:

```
#1 = 10000000;
```
(Assigns a value of 10^7 to #1)

```
#1 = #1 * #1 * #1 * #1 * #1 * #1;
```
(Assigns a value of 10^{42} to #1)

```
#1 = #1 * 100000;
```
(Assigns a value of 10^{47} to #1, which is the largest legal value)

```
#2 = #1 / 10 * 10;
```
(Sets #2 = 10^{47})

```
#3 = #1 * 10 / 10;
```
(Though the expression on the RHS is **mathematically the same** as the previous expression, the first calculation from the left gives 10^{48} which would cause data overflow, terminating the program execution. In such cases, brackets can be used to change the order of calculation)

```
#4 = #1 * [10 / 10];
```
(Sets #4 = 10^{47})

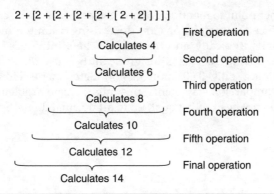

FIGURE 4.1 Nesting of brackets.

Nesting of Brackets

Nesting of brackets is permitted up to a maximum of **five** levels (i.e., five left and five right nested brackets can be used), including the brackets used with functions such as

SIN [<a variable number or an arithmetic value or an arithmetic expression>]

SQRT [<a variable number or an arithmetic value or an arithmetic expression>]

The calculation starts from the innermost brackets, and proceeds to the outermost brackets, one by one. The example given in Fig. 4.1 is trivial, but it does show the maximum possible level of nesting, and explains the order of calculation.

There is no limitation on the total number of brackets in an expression. The same expression can have nesting at several places, each nesting being restricted to a maximum of five levels.

4.2 Arithmetic Operations

The four arithmetic operators, addition (+), subtraction (−), division (/), and multiplication (*), carry usual meanings and give calculator-type results. In fact, these have already been used in several examples. The only thing that needs to be repeated here is that the control does not differentiate between real numbers and integer numbers. In fact, all numbers are treated as real numbers, in calculations. So,

```
#1 = 1 / 2;
#1 = 1.0 / 2;
#1 = 1 / 2.0;
#1 = 1.0 / 2.0;
```

are all equivalent, and assign a value of 0.5 to #1.

Another point to mention is about the accuracy of the calculated values. Since the calculations are done using a limited number of binary digits, typically an error of 10^{-10} gets introduced in **each** arithmetic operation. This is insignificant because the accuracy of the machine is only 0.001 mm in millimeter mode. However, in equality comparison statements, this might create problems, as the values may not be exactly equal, up to **all** decimal digits. So, instead of using

```
IF [#1 EQ #2] THEN #3 = 10;
```

use

```
IF [ABS[#1 - #2] LT 0.001] THEN #3 = 10;
```

assuming 0.001 is the acceptable error (or else, choose a still smaller value). Conditional statements are explained in detail in Chap. 5.

Division versus Block-Skip Function

The slash symbol (/) is used both for division and the *block-skip* function. Normally, a slash in the beginning of a block is interpreted as the block-skip symbol, and that in the middle of a block is taken as the division operator. However, some controls (including Fanuc) allow mid-block skip also. This may give unexpected results. So, on such controls, enclose the division operation within brackets:

```
#1 = 10;
#2 = #1 / 2;          (The slash may be interpreted as the block-skip
                       symbol. So, #2 is set to 10, if the block-skip switch
                       on the MOP is ON)
#3 = [#1 / 2];        (#3 is always set to 5, even if mid-block skip is per-
                       mitted and the block-skip switch is ON)
```

Some machine tool models do not have block-skip switches. On such machines, slash is always interpreted as the division operator; as such, there is no need to use brackets for division.

4.3   Trigonometric Functions

The available functions are

SIN
COS
TAN
ASIN
ACOS
ATAN

representing *sine, cosine, tangent, arc sine* (sin^{-1}), *arc cosine* (cos^{-1}), and *arc tangent* (tan^{-1}), respectively, the same as those found on scientific calculators. All inputs for SIN, COS, and TAN are in degrees, and the outputs of the inverse functions ASIN, ACOS, and ATAN are also in degrees. So, if an angle is given in degree/minute/second format, it would need to be converted into degrees using the conversion formula

$$D_d = D + \frac{M}{60} + \frac{S}{3600}$$

where D, M, and S are degree, minute, and second values of the specified angle, respectively, and D_d is the converted value in degrees.

Example:
```
10° 30′ 36″ = 10 + 0.5 + 0.01 = 10.51°
```

Like all other functions, the trigonometric functions also must be used with square brackets. The brackets may contain a constant, a variable number, or an arithmetic expression. Note that ATAN uses a rather unusual way of representation—the two sides (perpendicular and base) of the right-angled triangle are required to be specified within separate brackets, separated by a slash:

`#1 = SIN[30];`	(Sets #1 = 0.500)
`#2 = ACOS[#1];`	(Sets #2 = 60.000)
`#3 = TAN[#2 * 3 / 4];`	(Sets #3 = 1.000)
`#4 = ATAN[SQRT[3]] / [#1 * 2];`	(Equivalent to $tan^{-1}\sqrt{3}$. So, sets #4 = 60.000)
`#5 = TAN[90];`	(Overflow error, since tan $90° = \infty$)

Note that, like other macro functions, the trigonometric functions also are accurate up to about **eight** decimal digits. For example, ACOS[0.5] actually returns 59.999999 on Fanuc 0i control. Such inaccuracies may be found on all trigonometric functions, even if not explicitly mentioned in the examples given here.

The representation of ATAN is a unique feature of macro programming, which always gives the correct value of the angle, even if it is not an acute angle (ASIN and ACOS are mainly useful for acute angles):

`#5 = ATAN[1] / [1];`	(Sets #5 = 45.000)
`#6 = ATAN[1] / [-1];`	(Sets #6 = 135.000)
`#7 = ATAN[-1] / [-1];`	(Sets #7 = 225.000)
`#8 = ATAN[-1] / [1];`	(Sets #8 = 315.000)
`#9 = ATAN[1];`	(Syntax error)

When parameter 6004#0 is set to 0, the solution range of ATAN is 0° to 360°, as in the previous examples. When 6004#0 is set to 1, the range becomes –180° to 180° (note that the angle remains same, only its representation changes):

```
#10 = ATAN[1] / [1];        (Sets #10 = 45.000)
#11 = ATAN[1] / [-1];       (Sets #11 = 135.000)
#12 = ATAN[-1] / [-1];      (Sets #12 = -135.000)
#13 = ATAN[-1] / [1];       (Sets #13 = -45.000)
```

Like other functions, there is no need to enclose the ATAN function with outer brackets in arithmetic calculations:

```
#14 = ATAN[1] / [1] / [1 + 1];     (Sets #14 = 22.500)
```

Parameter 6004#0 does not affect ASIN or ACOS. Irrespective of the 6004#0 setting, the solution range of ASIN is 270° to 360° and 0° to 90°, whereas the range of ACOS is 0° to 180°. Hence, ASIN would never give an answer in the second or third quadrant. Similarly, ACOS does not give an answer in the third or fourth quadrant:

```
#15 = ASIN[0.5];      (Sets #15 = 30.000)
#16 = ASIN[-0.5];     (Sets #16 = 330.000)
#17 = ACOS[0.5];      (Sets #17 = 60.000)
#18 = ACOS[-0.5];     (Sets #18 = 120.000)
```

So, to avoid any confusion resulting in an inappropriate answer, use of ATAN is recommended whenever the angle is likely to vary in the entire range of 0° to 360°.

The magnitude of the value (i.e., the absolute value) specified for ACOS and ASIN must be less than or equal to 1, otherwise it would generate an alarm condition, terminating the program execution.

The angles for SIN, COS, and TAN can be positive or negative and can even be greater than 360°. Practically, there is no upper limit for the angle. Fanuc 0i control allows a maximum value of 1048575.999°. The corresponding angle in the 0° to 360° range is calculated by subtracting multiples of 360 from the specified angle. For example,

$$1048575.999 = 2912 \times 360 + 255.999$$

which is equivalent to 255.999°.

Since the accuracy of the trigonometric functions is typically 10^{-8}, a calculation such as SIN[0] does not return 0. Some typical calculation results are

```
SIN[0]     (Returns -0.46566129 × 10⁻⁸)
COS[90]    (Returns 0.37252903 × 10⁻⁸)
TAN[0]     (Returns -0.46566128 × 10⁻⁸)
```

Due to the limitation of the display screen, which can show only eight decimal digits and does not include any exponential digits, these values are displayed as *********, but used "correctly" in subsequent calculations:

```
1 / SIN[0]    (Returns 214748370.0)
```

Inaccuracy of this level is not important for the purpose of a CNC machine because the accuracy of the machine is typically 0.001 in millimeter mode and 0.0001 in inch mode. No matter how complex and extensive the calculation is, the resulting error is likely to be much smaller than the accuracy of the machine. However, as already suggested, the programmer must avoid direct equality comparison of, say, SIN[0] with zero. Instead, if required, the difference in values should be tested against a chosen small value:

`[SIN[0] EQ 0]`	(FALSE, so do not use this format)
`[ABS[SIN[0] - 0] LT 0.0000001]`	(TRUE, hence the recommended format)

As long as the programmer understands the implications of such inaccuracies, no problem would arise. However, parameter 6004#1 can be used to normalize the result to 0 whenever the calculated value of SIN, COS, and TAN is less than 10^{-8}. When 6004#1 is set to 1 (its default setting is 0),

`SIN[0]`	(Returns 0)
`COS[90]`	(Returns 0)
`TAN[0]`	(Returns 0)
`1 / SIN[0]`	(Execution terminates with an alarm message "DIVIDED BY ZERO")

4.4   Rounding Functions

Rounding is of two types: *implicit rounding* and *explicit rounding*. Implicit rounding is automatically done by the control, to suit the specific *address*. Several such cases have already been discussed in Chap. 2. Some similar examples are given here to refresh the memory (refer to Chap. 2 for details):

`#1 = SQRT[2];`	($\sqrt{2}$ = 1.41421356237. . . . The control uses only eight decimal digits, with adequate number of leading or trailing zeroes, to save a number. So, the value saved is 1.4142136)
`#1.5000000 = 1;`	(Equivalent to #2 = 1)
`M2.5 S1000.4;`	(Illegal statement)
`M[2.5] S[1000.4];`	(Equivalent to M03 S1000; rounding is done only if the number is enclosed by brackets)
`G[2.5];`	(Never use a G-code in this manner. It will either be an illegal command or may cause unexpected tool movement, causing accidents)

```
#3 = 2.4;
#4 = 2.04;
M#3;                    (Equivalent to M02)
G#3;                    (Illegal statement)
G#4;                    (Equivalent to G02)
#5 = 123.45678;
G00 X#5;                (Equivalent to G00 X123.457 in millimeter
                        mode)
```

As a good programming practice, **one should never rely on implicit rounding,** as this is highly error-prone. If the programmer does not clearly understand the logic built in the controller, unexpected machine behavior may result. Implicit rounding should only be done for the purpose of using a real number as an integer number, by **discarding zeroes after the decimal point** (e.g., 1.0 being converted to 1). For example, G#1 is equivalent to G01, if the value stored in #1 is 1, 1., 1.0, or 1.0000000. In fact, this becomes necessary because the control does not define a variable as an integer variable; **all variables store real numbers**.

ROUND, FIX, and FUP

Three functions are available for explicit rounding: ROUND, FIX, and FUP. These are used for rounding a real number to the nearest integer, the lower integer, and the upper integer, respectively. The operation of these functions is similar to that in a typical high-level programming language, except that the CNC control saves the integer answer as a real number with zero after the decimal point!

Examples:
```
ROUND[10.0]            (Returns 10.0)
ROUND[10.2]            (Returns 10.0)
ROUND[10.499999]      (Returns 10.0)
ROUND[10.5]            (Returns 11.0)
ROUND[10.8]            (Returns 11.0)
FIX[10.0]              (Returns 10.0)
FIX[10.2]              (Returns 10.0)
FIX[10.499999]        (Returns 10.0)
FIX[10.5]              (Returns 10.0)
FIX[10.8]              (Returns 10.0)
FUP[10.0]              (Returns 10.0)
FUP[10.000001]        (Returns 11.0)
FUP[10.2]              (Returns 11.0)
FUP[10.499999]        (Returns 11.0)
FUP[10.5]              (Returns 11.0)
FUP[10.8]              (Returns 11.0)
```

The result of operating these functions on a negative number is the same as that in the case of a positive number, except that the negative sign is retained:

ROUND[-10.2]	(Returns –10.0)
ROUND[-10.8]	(Returns –11.0)
FIX[-10.2]	(Returns –10.0)
FIX[-10.8]	(Returns –10.0)
FUP[-10.2]	(Returns –11.0)
FUP[-10.8]	(Returns –11.0)

These functions can also be applied on variables that store some value. The following examples assume that #1 stores 1.3, and #2 stores 1.7:

ROUND[#1]	(Returns 1.0)
ROUND[-#2]	(Returns –2.0)
FIX[#1]	(Returns 1.0)
FIX[-#2]	(Returns –1.0)
FUP[#1]	(Returns 2.0)
FUP[-#2]	(Returns –2.0)

It is also possible to round a value to the desired number of places after the decimal point, which may be needed for some special application. For example, if 100.12345 is required to be rounded up to four places after the decimal point, carry out the following procedure:

#1 = 100.12345;	
#2 = #1 * 10000;	(Sets #2 = 1001234.5)
#3 = ROUND[#2];	(Sets #3 = 1001235.0)
#4 = #3 / 10000;	(Sets #4 = 100.12350)
#1 = #4;	(#1 contains rounded value up to four places after the decimal point, 100.1235)

These commands are given separately for the purpose of explaining the intermediate calculation steps. It is also possible to combine these in a single nested block, in the following manner:

```
#1 = ROUND[#1 * 10000] / 10000;
```

For rounding to three places after the decimal point, use 1000 as the multiplier; for two places after the decimal point, use 100, and so on.

Special care is needed while using the FIX function because of a possible inaccuracy in macro calculations. For example, if variable #1 contains 0.002, and variable #2 is defined as #1 * 1000, the resulting value may not exactly be 2. It may typically be 1.9999999. As a result, FIX[#2] would return 1.0! So, in such cases, either use the ROUND function, or modify the calculated value appropriately. Since the least input increment of the machine in millimeter mode is 0.001, it is safe to add 0.0001 to the calculated value, for the purpose of rounding it down. So, instead of using FIX[#2], use FIX[#2 + 0.0001], which will solve the problem.

Finally, it is very important to understand that if ROUND is used for some **axis distance in an NC statement**, rounding is done up to the **least input increment** of the machine. The least input increment in millimeter mode is typically 0.001 mm, as assumed in the following example:

```
#1 = 100.12345;
G00 X[ROUND[#1]];       (Places the tool at X = 100.123)
```

In fact, the ROUND function is redundant here. As already explained in Chap. 2, even without the ROUND function, implicit rounding would be done up to the least input increment. So, the second statement is equivalent to

```
G00 X#1;
```

Also note the presence of the outer bracket in the ROUND function. If a function or an expression is being used as the value in an NC word, it must be enclosed in brackets is:

```
X ROUND[1.2];       (An illegal command)
X[ROUND[1.2]];      (Equivalent to X1.200)
```

It is not that the ROUND function is completely useless in an NC statement (as can be concluded from the given example). In fact, it can be very effectively used to eliminate the round-off errors during program execution. Consider the following case, with 0.001 as the least input increment, on a milling machine (assume that the current tool position is X100, along the X-axis):

```
#1 = 100.2345;
#2 = 200.3456;
G91 G00 X#1;        (Places the tool at X200.235)
X#2;                (Places the tool at X400.581, because 200.235 +
                    200.346 = 400.581)
X-[#1 + #2];        (Equivalent to X-[300.5801] which is rounded to
                    X-300.580. So, the tool is placed at X100.001, which
                    is not the initial position of the tool)
```

Though the difference between the expected position (X100) and the actual position (X100.001) is very small (0.001), which may not be of any practical concern, it will cause a **logical flaw** in an application involving comparison of tool positions. Moreover, since the error is accumulative, it is possible that, after several such operations, the tool may actually deviate from the correct position to an unacceptable extent. This problem can easily be obviated by replacing the last statement by

```
X-[ROUND[#1] + ROUND[#2]];
```

which is equivalent to X-[100.235 + 200.346], that is, X-300.581, which would place the tool **exactly** at the expected position.

Note that rounding of axis distances up to the least input increment is done only if the values are specified in terms of **variables**, as in the cases discussed above. **A constant value is truncated up to the least input increment.** For example, G91 G00 X100.2345 would move the tool by 100.234 mm.

4.5 Miscellaneous Functions

Additional functions for evaluating square root (SQRT), absolute value (ABS), natural logarithm (LN), and exponential value (EXP) are available. The calculated values are automatically rounded to eight decimal places. SQRT and ABS carry usual meanings and have already been used in different contexts.

SQRT

This calculates the square root of a given **positive** number (a negative number would give an "ILLEGAL ARGUMENT" alarm message). SQRT is the only available function in its category. There is no function available for calculating, say, the cube root or square of a number. However, as explained later, a combination of LN and EXP functions can be used for calculating any arbitrary exponent of a given number.

Examples:
```
SQRT[-2]    (Illegal argument)
SQRT[0]     (Returns 0.000)
SQRT[2]     (Returns 1.4142136)
```

ABS

This function simply changes the sign of a negative value, and has no effect on a positive value.

Examples:
```
ABS[-2]     (Returns 2.000)
ABS[0]      (Returns 0.000)
ABS[2]      (Returns 2.000)
```

The "innocent-looking" ABS function comes in very handy at times; some such examples have already been discussed. As another example, consider a macro that involves a drilling operation using G81, on a milling machine. The usual choice for Z0 level is the top surface of the workpiece. Then, the depth of hole must have a negative sign in the G81 block. So, when a program calls this macro, a negative value for the depth of the hole must be passed on to the macro (the method of passing on desired values to the local variables of a macro has not been formally discussed so far. So, without bothering about the exact procedure, just try to understand the theme of the discussion). If, by mistake or due to a misunderstanding, a programmer specifies a positive value for the depth, G81 will start "drilling" in the positive Z-direction, away from the workpiece! To make sure that **both** positive and negative values work equally well, it is necessary that G81 uses the negative of the absolute value of the specified depth. This will ensure that the Z-value for G81 is always negative. For example, if variable #26 is used for passing on the value for the depth of the hole, the macro should use G81 in the following manner:

```
G81 X_ Y_ Z-[ABS[#26]] R_ K_ F_;
```

Since a macro is supposed to be a general-purpose program that is used (called) by several programs written by different programmers, such inconsistencies are not very uncommon. Developing a macro requires an in-depth knowledge of macro programming, which is a rather vast and specialized subject. So, most of the programmers have to use the macros developed by experts, as a black box. Writing a macro might be difficult, but, as we will see later, using a macro is as simple as using a built-in canned cycle. One only needs to know the purpose of the different letter-addresses of the macro. Though it is expected that a macro programmer would properly document the macro developed by him, by inserting suitable comments in the macro and/or by explaining its use on separate handouts, care has to be taken for common mistakes. A professional approach to macro programming aims at **anticipating what can go wrong, before it actually goes wrong!**

LN

This function calculates the *natural* logarithm, that is, the logarithm to the base e (2.7182818). The logarithm to the base 10 (i.e., LOG function) is not available. If required, it can be calculated using the formula

```
log x = (ln x) / (ln 10), that is, LN[x] / LN[10]
```

A negative or zero value for the argument of LN is invalid, which would give a "CALCULATED DATA OVERFLOW" alarm message.

Examples:
```
LN[0]     (Results in overflow error)
LN[1]     (Returns 0.000)
LN[2]     (Returns 0.6931472)
```

EXP

This function calculates the antilog of the natural logarithm (i.e., e^x). The maximum permissible value for its argument is about 110.2. A higher value would give "CALCULATED DATA OVERFLOW," as the calculated value would approach/exceed 10^{48}. Antilog to the base 10 (i.e., 10^x) is not available. If required, it can be calculated using the formula

```
10ˣ = eˣ ˡⁿ ¹⁰, that is, EXP[x * LN[10]]
```

Examples:

EXP[-2]	(Returns 0.1353353)
EXP[0]	(Returns 1.000)
EXP[2]	(Returns 7.3890561)
EXP[110.25]	(Overflow error)

Arbitrary Exponent of a Number

No standard function is available to evaluate an expression in the form x^y. It can, however, be calculated in an indirect manner, using the formula

$x^y = e^{y \ln x}$, that is, EXP[y * LN[x]]

Examples:

EXP[4 * LN[3]]	(Equivalent to 3^4, which returns 81.000)
EXP[4 * LN[10]]	(Equivalent to 10^4, which returns 10000.000)

4.6 Logical Functions

The available logical functions are AND, OR, and XOR. These can be used in two different manners, as

- Bitwise functions
- Boolean functions

Bitwise Functions

When used as bitwise functions, all the corresponding bits of the binary representations of the two numbers (on which these function are operated) are individually compared, and the resulting answer in the binary format is converted back to the decimal format. Table 4.1 is the truth table for the three functions. Bit-1 and bit-2 are the **corresponding bits** of the two numbers.

Figure 4.2 gives the Venn diagram representations of these functions. In this figure, the circular area corresponding to a bit represents a binary value of 1, and the area lying outside the circle represents binary value 0. The shaded area represents the result (binary value 1) of the specified logical operation. This figure also indicates that all the

Bit-1	Bit-2	AND	OR	XOR
0	0	0	0	0
0	1	0	1	1
1	0	0	1	1
1	1	1	1	0

TABLE **4.1** Truth Table for Bitwise AND, OR, and XOR

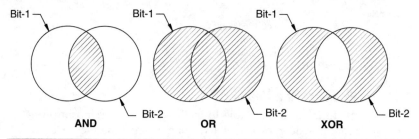

FIGURE 4.2   Venn diagram representation for bitwise AND, OR, and XOR.

three logical operations are associative. So, for example, #1 AND #2 is equivalent to #2 AND #1.

Some examples of bitwise logical operations are given below:

```
#1 = 4;              (Same as binary 0100)
#2 = 3;              (Same as binary 0011)
#3 = 1;              (Same as binary 0001)
#4 = #1 AND #2;      (Result in binary form is 0000. So, #4 is set to 0.000)
#5 = #2 AND #3;      (Result in binary form is 0001. So, #5 is set to 1.000)
#6 = #1 OR #2;       (Result in binary form is 0111. So, #6 is set to 7.000)
#7 = #2 OR #3;       (Result in binary form is 0011. So, #7 is set to 3.000)
#8 = #1 XOR #2;      (Result in binary form is 0111. So, #8 is set to 7.000)
#9 = #2 XOR #3;      (Result in binary form is 0010. So, #9 is set to 2.000)
```

The logical functions are operated upon **integers**. If an integer value (or a variable containing an integer value) is not specified, rounding is automatically done to the **nearest** integer:

```
#1 = 1.5;
#2 = 1.4999999;
#3 = #1 AND #2;       (Equivalent to #3 = 2 AND 1; which is
                       bitwise AND of binary numbers 010
                       and 001. The result is binary 000, which
                       is the same as decimal 0. So, 0.000 is
                       stored in #3)

#4 = 2.5 XOR 2.4999999;  (Equivalent to #4 = 3 XOR 2; which is
                       bitwise XOR of binary numbers 011
                       and 010. The result is binary 001, which
                       is the same as decimal 1. So, 1.000 is
                       stored in #4)
```

Though the bitwise operations are actually logical operations, they look similar to arithmetic operations. In fact, the result of bitwise operations can be used in arithmetic calculations also, though their purpose is entirely different. So, a statement like

```
#1 = #2 * #3 AND #4;
```

may be meaningless, but not illegal. Note that since multiplication and AND have equal priority, the expression would be evaluated from left to right as

```
#1 = [#2 * #3] AND #4;
```

On the other hand, multiplication has higher priority than OR. Hence, both the following statements are equivalent:

```
#1 = #2 OR #3 * #4;
#1 = #2 OR [#3 * #4];
```

If evaluation of OR is desired first, brackets will have to be used:

```
#1 = [#2 OR #3] * #4;
```

One may not always remember all the priority rules, so it is a good practice to use brackets to avoid confusion. Extra brackets do no harm, except that the maximum permitted level of nesting is five. If this limitation creates problem, or if multiple nesting makes the expression too complex, break the expression into smaller parts, spread over multiple lines.

Boolean Functions

The logical AND, OR, and XOR can also be used as Boolean functions, operated upon two Boolean expressions, and the result of operation is a Boolean value—TRUE or FALSE (recall that bitwise operation is done on two arithmetic values or arithmetic expressions, and the result of operation is an arithmetic value). These are often used in IF and WHILE conditional statements (conditional statements are described in detail in the next chapter), in a manner such as

```
IF [<conditional expression-1> AND <conditional
expression-2>] THEN #1 = 1;
```

which sets #1 = 1, if both conditions are TRUE.

Such a use of AND, OR, and XOR is similar to AND, OR, and XOR functions of any high-level computer programming language, and the meaning is the same as that in plain English. The truth table (given in Table 4.2) is similar to that of bitwise functions.

Condition-1	Condition-2	AND	OR	XOR
FALSE	FALSE	FALSE	FALSE	FALSE
FALSE	TRUE	FALSE	TRUE	TRUE
TRUE	FALSE	FALSE	TRUE	TRUE
TRUE	TRUE	TRUE	TRUE	FALSE

TABLE 4.2 Truth Table for Boolean AND, OR, and XOR

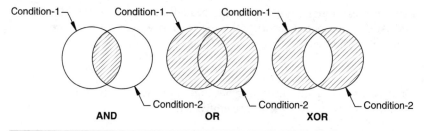

FigURE **4.3**   Venn diagram representation for Boolean AND, OR, and XOR.

The Venn diagram representation (shown in Fig. 4.3) also is similar to that for bitwise functions. In this figure, the circular area corresponding to a condition represents a Boolean value of TRUE, and the
area lying outside the circle represents a Boolean value of FALSE. The
shaded area represents the result (Boolean value TRUE) of the specified logical operation.

It is important to note that **a Boolean value TRUE is not equivalent to the arithmetic value 1,** and similarly, **FALSE is not equivalent
to 0.** So, a statement such as

```
#1 = #2 EQ #3; or
#1 = [#2 EQ #3];
```

is meaningless, and would give "FORMAT ERROR IN MACRO." It is
not possible to assign a Boolean value (TRUE or FALSE) to a variable,
which can only be assigned an arithmetic value. Confusion prevails
among some macro programmers that if a Boolean value is assigned
to some variable (as in the given example), 1 or 0 gets stored in the
variable, corresponding to TRUE and FALSE, respectively.

Bitwise versus Boolean Operations

By now, the reader must have realized that whenever AND, OR, and
XOR are operated upon **arithmetic values**, these functions behave as
bitwise functions, and the result is an **arithmetic value.** On the other
hand, if these functions are operated upon **Boolean values,** these
become **Boolean functions,** returning a **Boolean value** (TRUE or
FALSE). And, of course, **arithmetic and Boolean values cannot be
mixed up with AND, OR, and XOR:**

`1 LT 2 AND 3`	(Presence of conditional operators, such as EQ and LT, makes an expression a Boolean expression. A Boolean expression such as this is evaluated from left to right, if there are no brackets. Hence, this expression is equivalent to TRUE AND 3, which is meaningless, hence illegal)
`[1 LT 2] AND 3`	(Same as above, hence illegal)

`1 LT [2 AND 3]`	(Equivalent to 1 LT 2, which evaluates to TRUE)
`1 AND 2 LT 3`	(Equivalent to 0 LT 3, which evaluates to TRUE)
`1 AND [2 LT 3]`	(Equivalent to 1 AND TRUE, which is meaningless, hence illegal)
`[1 LT 2] AND [3 LT 4]`	(Equivalent to TRUE AND TRUE. Hence, returns TRUE)

Similarly, Boolean values cannot be used in any arithmetic operation:

`1 * [2 LT 3]`	(Equivalent to 1 * TRUE, which is meaningless, hence illegal)
`1 * 2 LT 3`	(Equivalent to 2 LT 3, which evaluates to TRUE)

Skipping brackets and relying exclusively on priority rules may make a program difficult to interpret in certain cases, some of which have already been discussed. A program should not just be error-free; it is important that its logic also is easy to follow. So, especially in complex conditional statements, use of extra brackets is highly recommended. Priority rules should mainly be used for arithmetic operations only. And, as a rule of thumb, a conditional expression must be enclosed within brackets.

Enabling Boolean Operations

AND, OR, and XOR, as bitwise functions, are always available on the machine. However, their availability as Boolean functions is parameter (6006#0) dependent. The default setting of 6006#0 is 0, which does not allow AND, OR, and XOR to be used as Boolean functions. If their use as Boolean functions is also desired, set this bit to 1. With this setting, these functions can be used both as bitwise functions and Boolean functions. An example of a mixed use is given below, where behavior (bitwise or Boolean operation) of the AND function depends on **the context of its use**:

```
IF [[[#1 AND #2] LT 2] AND [#3 LT #4]] THEN #5 = 1;
```

Here, the first (from left) occurrence of AND is a bitwise operation, whereas the second occurrence is a Boolean operation.

An Application Example of Bitwise Operation

The Boolean AND, OR, and XOR are used more often than bitwise AND, OR, and XOR in common applications. Bitwise functions are normally used in conjunction with system variables for interface signals for interacting with external devices connected to the CNC.

As an example, consider the case where several pallets are available for holding the workpieces for a milling machine, and the program needs to know whether the correct pallet is loaded. Each pallet has a unique designation number (e.g., 27), and a binary coding system to indicate this number. It also has a number of independent N/O (*normally open*) limit switches (one switch for each binary signal) to indicate whether it is loaded or empty.

Though the method described below is simple, it is not the best method, given practical considerations. It is given here because unless one realizes the limitation of a particular solution, one will not be able to appreciate the best solution. A better method is described at the end.

In a simple coding arrangement, there can be an eight-pole *dual in-line package (DIP)* switch attached to each pallet (assuming, the largest pallet numbers is 255 or less), and the switch setting would indicate the pallet number (in binary, i.e., 0 for OFF and 1 for ON), which would be 00011011 corresponding to 27. A DIP switch is a set of manual electric switches that are arranged in a group in a standard package. The whole package unit is referred to as a DIP switch in the singular. DIP is sometimes also referred to as DIL (*dual in-line*). This type of switch is designed to be used on a *printed circuit board* (PCB) along with other electronic components and is commonly used to customize the behavior of an electronic device for specific situations. Several types of DIP switches are available. The slide-type and the rocker-type DIP switches are very commonly used for binary coding. These are arrays of simple *single-pole, single-throw* (SPST) contacts, which can be either ON or OFF. This allows each switch to select a one-bit binary value. The values of all switches in the package can be interpreted as a decimal number. Eight switches offer 256 combinations—0 to 255—corresponding to 00000000 and 11111111, respectively. A common multipin IC is another example of a dual in-line package.

Each pole of the DIP switch is connected in series with an independent N/O limit switch, so that the binary signals would be sent only if the limit switch is pressed. A physical connection, through a ribbon cable (having eight signal wires and one common wire) between the terminal strip of the input/output (I/O) module of the PMC and the limit switches of each pallet, would be needed for sending the signals to the PMC.

Addition of eight rungs in the PMC ladder, one for each input signal, for writing these signals (in a sequential manner) to the appropriate G-signals (G54.0 to G54.7), would define the corresponding system variables (#1000 to #1007, and #1032). While #1000 to #1007 would contain 0 or 1, corresponding to the OFF/ON state of the corresponding switch, #1032 would contain the complete information in its 16 bits (in fact, the decimal interpretation of the binary representation, 27 in this case, would get stored in #1032). So, by analyzing the content of #1032 bitwise AND/OR/XOR are needed for this purpose. For example, #1032 XOR 27 would return 0 for #1032 = 27, and a nonzero number otherwise. So, a 0 value would confirm that pallet number 27 is loaded), information about the loaded pallet becomes available inside the program that may take appropriate machining decisions, suitable for the workpiece on that specific pallet.

The method explained above is theoretically correct, but not practical, because of a possible interference caused by the cables (that run from each pallet to the terminal strip) in a moving system. A practical method is described below.

First determine the maximum number of different pallets there will be. Two pallets can be encoded with just one switch (ON or OFF). Four pallets with two switches (OFF/OFF, ON/OFF, OFF/ON, and ON/ON), eight pallets with three switches, sixteen pallets with four switches, and so on. There is no need to use eight switches unless there are 256 pallets. In addition to the "binary number" switch array, an additional switch would be needed to confirm whether the pallet is loaded and locked. Now, what remains is how to actuate these switches.

All the switches (push-to-ON type) are mounted on the stationary part of the machine, underneath where the pallets are clamped down. The pallets simply have protruded (meaning 1) or recessed (meaning 0) little buttons (the number of buttons being equal to the number of switches) on the bottom, that line up with the switches mounted on the machine. When a pallet is locked down, a certain combination of switches gets actuated by the buttons on the pallet, indicating its number. The additional switch, to sense whether or not a pallet is loaded, is actuated by all the pallets in the same manner. And there is no need to have a physical series connection between this switch and the switches used for pallet coding purpose (in fact, for such a connection, several independent switches would be needed—one for each coding switch). Instead, it may appear as an N/O relay, in series, in all the ladder rungs corresponding to other switches. The array of switches and the additional switch are permanently wired to the I/O module with a cable. So, just a single cable is needed, which always remains stationary, fixed to the stationary parts of the machine.

A more sophisticated pallet system might use bar-coding or a radio frequency (RF) tag to pass more information to the machine than just a pallet number.

Knowledge of the ladder language is a prerequisite for such applications. Though ladder language is not covered in this text, a brief description of wiring methods is given in Chap. 12.

4.7 Conversion Functions

The available functions are BIN and BCD which are used for converting a number from binary-coded decimal format to binary format, and vice versa. These functions are used for two-way signal exchange between the PMC (which uses signals in binary format) and the

Decimal Number	BCD Representation	Binary Interpretation
0	0000 0000 0000 0000	0
1	0000 0000 0000 0001	1
9	0000 0000 0000 1001	9
10	0000 0000 0001 0000	16
19	0000 0000 0001 1001	25
20	0000 0000 0010 0000	32
99	0000 0000 1001 1001	153
100	0000 0001 0000 0000	256
999	0000 1001 1001 1001	2457
1000	0001 0000 0000 0000	4096
9999	1001 1001 1001 1001	39321

TABLE 4.3   BCD Representation of Decimal Numbers

external devices (that might be using signals in binary-coded decimal format, e.g., for seven-segment display of each digit of a number, separately), connected to the CNC.

Recall that, in the binary-coded decimal (BCD) representation of a decimal number, each digit of the number is **independently** represented by **four binary digits**. Refer to Table 4.3 for some examples of BCD representations of up to four-digit decimal numbers. Note that four binary digits can have a highest decimal value of 15, but since the highest decimal digit is only 9, 1010 and above (i.e., up to 1111) do not appear in BCD representation.

Format:

```
#i = BCD[<binary interpretation of a binary number>];
#i = BIN[<binary interpretation of a BCD number>];
```

The arguments of both BCD and BIN are **binary interpretations**, and the results of conversion are again **interpreted in binary**. Note that the binary interpretation of a binary number is the actual number, but such an interpretation of a BCD number is not the number that the BCD represents. **Format conversion is done with respect to the actual decimal number that is being represented in the two formats.**

Examples:

```
#1 = BCD[10];
```
(Converts 1010 to 0001 0000, corresponding to equivalent decimal number 10. Explanation: There is a binary number (1010) which, when interpreted in binary, is 10. So, 10 is the argument of the BCD function. This number,

when converted to BCD format, becomes 0001 0000. The decimal interpretation of the converted BCD representation is 16. So, #1 stores 16)

#2 = BIN[16]; (Converts 0001 0000 to 1010, corresponding to equivalent decimal number 10. Explanation: There is a BCD number (0001 0000), which, when interpreted in binary, is 16. So, 16 is the argument of the BIN function. This number, which is actually decimal 10, becomes 1010 when converted to binary format. The decimal interpretation of the converted binary representation is 10. So, #2 stores 10)

Note that, in both the examples, the same decimal number 10 is involved in conversion. The concept would be easier to understand if it is realized that **the representation of a given decimal number in one format is converted to the representation of the same decimal number in the other format.**

Following similar logic, verify, as an exercise, that BCD[64] returns 100, and BIN[100] returns 64:

0110 0100 ↔ 1000000 (Both represent decimal number 64 in BCD and binary formats, respectively)

Negative values for the arguments of BCD and BIN are not allowed (an "ILLEGAL ARGUMENT" alarm message would be displayed). The arguments should be integers. If a real value is specified, rounding to the nearest integer is automatically done.

These functions are used in conjunction with system variables for interface signals (explained in Chap. 3), which are used for receiving/ sending 16-bit/32-bit signals to/from the PMC. For example, if variable #1032 corresponds to input status 0000000000010011, it will store a decimal value of 19.000:

#1 = #1032; (#1 is set to 19.000)

If the same signal is to be sent to an external device(s), in the BCD format, through variable #1132, BCD function would be needed:

#2 = BCD[#1]; (The BCD representation of 19 is 0000 0000 0001 1001. Its binary interpretation is 25. So, #2 is set to 25)

#1132 = #2; (#1132 sends the required BCD signal, 0000000000011001)

In fact, there is no need to involve intermediate variables. The following statement would do the same thing:

#1132 = BCD[#1032];

The BIN and BCD functions can also be used for variable #1133 that sends a 32-bit signal. Since the representation of a single-bit signal, in both binary and BCD formats, is the same, these functions are not needed for variables #1000 to #1015, and #1100 to #1115.

CHAPTER 5

Branches and Loops

In a program, the sequence of execution can be changed using GOTO and IF_GOTO_ statements, which is called *branching*. These statements can also be used to create a *loop*, and execute it repeatedly until a certain condition gets satisfied. This, however, is not considered a good method of creating loops. It is a standard practice to avoid the use of GOTO in any structured programming language (such as Pascal), as it makes the program less readable. Moreover, it also introduces an element of risk, as a subsequent change in the sequence numbers might make the program unusable. Better methods are available. Custom Macro B provides the WHILE statement for this purpose.

5.1 Unconditional Branching

The format is

```
GOTO n;
```

where n is the desired *sequence number* (1 to 99999).

This causes an unconditional jump to the specified sequence number (recall that sequence number is also referred to as block number or line number or N-number). The sequence number in the GOTO statement can also be specified using a variable or an expression.

If the specified sequence number does not lie in the range 1 to 99999, or if the program does not contain the specified sequence number, further execution of the program terminates with an alarm.

It is not necessary to have sequence numbers for all the blocks of the program; only the target blocks must have sequence numbers.

Duplication of sequence numbers is permitted in a part program, though not recommended, as it might cause inadvertent errors. With the GOTO statement, it would result in an ambiguous branching instruction, generating an error condition. So, for the sake of good programming practice, the same sequence number must not appear more than once in a part program, especially when the program uses macro programming features.

Some examples of unconditional jump are given below:

```
GOTO 100;              (Flow of program execution jumps to N100 block. If
                       such a sequence number does not exist in the pro-
                       gram, further execution would be terminated, and an
                       alarm message would be displayed on the screen)
#1 = 1000;
GOTO #1;               (Jumps to N1000 block)
GOTO[#1 + 100];        (Jumps to N1100 block)
GOTO 0;                (Illegal command)
GOTO 100000;           (Illegal command)
```

The argument of GOTO is expected to be an integer. However, if a real number is specified, or if the variable or the expression, used as the argument, evaluates to a real number, rounding is automatically done to the nearest integer:

```
GOTO 100.0;            (Equivalent to GOTO 100)
GOTO 100.49999;        (Equivalent to GOTO 100. More than eight dig-
                       its, such as 100.499990, is not allowed)
GOTO 100.5;            (Equivalent to GOTO 101)
#1 = 100.49999;
GOTO #1;               (Jumps to N100 block)
GOTO[#1 + .00001];     (Jumps to N101 block)
```

Though such a use of GOTO is only a theoretical possibility, and never likely to occur in a practical situation, a thorough understanding of the logic being followed by the control helps in error diagnosis. At least, typing mistakes are always possible, which might give unexpected results, without generating any error message!

5.2 Conditional Branching

In conditional branching, GOTO is used in conjunction with a conditional expression. The format is

```
IF [<a conditional expression>] GOTO n;
```

If the specified condition is satisfied (i.e., evaluates to TRUE), the program execution jumps to sequence number n. If the specified condition is not satisfied (i.e., evaluates to FALSE), execution proceeds to the next block. The conditional expression must be enclosed within square brackets.

The six available conditional operators, which can be used in a conditional expression, carry the meanings given in Table 5.1 (these operators have already been used in various contexts; here they are described formally). Note that the equivalent mathematical symbols, shown in this table, are given for reference only. These cannot be used as conditional operators.

Operator	Meaning	Mathematical Symbol
LT	Less than	<
LE	Less than or equal to	≤
EQ	Equal to	=
NE	Not equal to	≠
GE	Greater than or equal to	≥
GT	Greater than	>

TABLE **5.1**   Boolean or Conditional Operators

The conditional operators are used for comparison between two constants/variables/arithmetic expressions, in any combination (e.g., comparison between a constant and a variable, or between a variable and an arithmetic expression, etc). The arithmetic expression(s), if any, must be enclosed within separate square brackets.

Examples:
```
[2 LT 3]
[2 LT #1]
[#1 LT #2]
[2 LT [#2 + #3]]
[#1 LT [#2 * 5]]
[[#1 * #2] LT [#3 + 10]]
```

The arithmetic expressions can also use any of the available mathematical functions and/or bitwise logical functions:

```
[1 LT TAN[#1]]
[#1 LT [#2 AND #3]]
```

An arithmetic expression can have multiple levels of nesting. But overall, there can be up to a maximum of five left and five right **nested** brackets. In a complex expression, always count the number of brackets to make sure that the left and the right brackets are equal in number, otherwise, the expression would become meaningless, generating an alarm condition:

```
[#1 LT [ROUND[SQRT[#2 * [#3 + #4]]]]]
```

This expression has five left and five right brackets. Note that the two consecutive functions (ROUND and SQRT, in this example) must be separated by square brackets.

Too complex an expression should preferably be broken into smaller parts, spread over several lines, to make it easy to interpret.

For example, the expression given above may be replaced by the following blocks:

```
#100 = #2 * [#3 + #4];
#100 = SQRT[#100];
#100 = ROUND[#100];
```

Thereafter, a simple expression [#1 LT #100] can be used in place of the original expression.

It is also permitted to use a combination of multiple conditions, coupled with Boolean AND, OR, and XOR, **provided the parameter 6006#0 is set to 1**. For example, it may be desired to take an action if two conditions are **both** TRUE at the same time. Such an example is already given toward the end of Sec. 4.6. In fact, any combination of the three Boolean functions can be used to simulate a complex condition. Note that, in a complex expression, it is better to make use of brackets to indicate the hierarchy of evaluation, even if these may actually be redundant due to priority rules. This makes the expression more readable and reduces the possibility of logical errors:

```
IF [[[#1 LT #2] AND [#3 LT #4]] OR [[#5 LT #6] XOR [#7
LT #8]]] GOTO 100;
```

The total number of nested left and right brackets is three each in this command. Without nesting, any number of brackets can be used. Seven left and seven right brackets have been used in this example.

Finally, as a practical example of conditional branching, consider the problem of calculating the sum of all numbers from 1 to 10 (i.e., 1 + 2 + 3 + ⋯ + 9 + 10). The flowchart for a summation algorithm, along with its execution trace, is shown in Fig. 5.1. It is a good programming practice to first prepare a flowchart for the chosen algorithm, and trace the execution, as far as possible. Program coding should be started only after fully verifying the algorithm. This reduces the possibility of logical errors to a great extent.

There is one more advantage of working with a flowchart. The macro variables do not reflect the physical meaning of a variable (which is really a serious limitation of macro programming). For example, if variable #100 is defined to contain the value of a certain variable, one would always have to remember the meaning of #100. This becomes difficult when working with a large number of variables. As a result, developing a program becomes a bit difficult because one does not have an intuitive feeling of which variable is being used for which purpose. The human brain does not work like a computer. In fact, this is one of the reasons why some programmers find macro programming too complex. A flowchart solves this problem. Once it is prepared, one simply has to copy it in terms of macro variables, and the program is ready! So, drawing a flowchart should not be considered an additional and unnecessary exercise. It does take some time

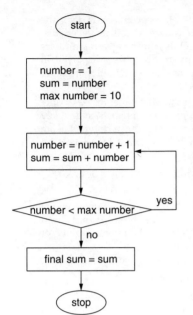

The table in the figure:

execution trace	number	sum
initial values of number and sum	1	1
values after executing I loop	2	3
values after executing II loop	3	6
values after executing III loop	4	10
values after executing IV loop	5	15
values after executing V loop	6	21
values after executing VI loop	7	28
values after executing VII loop	8	36
values after executing VIII loop	9	45
values after executing IX loop	10	55
final sum		55

FIGURE 5.1   Flowchart and execution trace of a summation algorithm.

initially, but ultimately, a complex program may be ready in much less time. Moreover, in a complex program, error diagnosis becomes extremely difficult if there is no flowchart. In the macro programming area, many a time, the slower one moves, the faster one reaches the destination. So, as a rule of thumb, one should never start program coding without first preparing a flowchart, especially if one is new to macro programming. Otherwise, the abstract nature of a macro program, which makes it look too complex, may discourage a new learner to study it further, and the battle would be lost before fighting it!

The next step after algorithm verification is to assign variable numbers to all the variables which appear in the flowchart. For the sake of proper documentation, the chosen variables should be described in the beginning of the program, within comments, for the benefit of other users who may wish to analyze the program. Now, writing the program is simply a matter of converting the flowchart into a coded language. Program number 8000 is based on the flowchart given in Fig. 5.1.

```
O8000 (SUMMATION OF NUMBERS);
```

(The program number for a macro is usually chosen from 8000 or 9000 series, as it is possible to edit-protect these programs, to avoid their accidental editing/deletion. Parameter settings 3202#0 = 1 and 3202#4 = 1 edit-protect program numbers 8000 to 8999, and

	9000 to 9999, respectively. In fact, the protected programs are not even displayed. Only the program numbers of such programs can be displayed in directory search, provided parameter 3202#6 is set to 1)
(#1 @ NUMBER);	(Note that the MDI panel has keys only for the upper-case alphabets, and no key for the @ character. @ can be typed using a soft key. Refer to Sec. 2.2 for details)
(#2 @ SUM);	
(#3 @ MAX NUMBER);	(Up to which summation is desired)
(#500 @ FINAL SUM);	
;	(This is a blank NC statement, which does not do anything. It has been inserted here for better readability of the program)
#1 = 1;	(Initial value of NUMBER)
#2 = #1;	(Initial value of SUM)
#3 = 10;	(Summation from 1 to 10 is desired)
N10 #1 = #1 + 1;	(NUMBER incremented by 1)
#2 = #2 + #1;	(SUM calculated)
IF [#1 LT #3] GOTO 10;	(Branching to N10 if the current value of the NUMBER is less than 10)
#500 = #2;	(The final value of the SUM stored in the permanent common variable #500)
M30;	(Replace M30 by M99 if this program is to be called by other programs)

Note that M02 and M30 include RESET action. So, these clear all local variables (i.e., #1 to #33) as well as common variables (#100 to #199) to null. Hence, if a value is required to be available even after the end of a program (e.g., for examining its value or for using it across **different** programs), it has to be stored in a permanent common variable. So, when the execution of the given program ends, #1, #2, and #3 are cleared to null, and the final value of the sum (which is 55, in this example) remains stored in #500, which can be displayed on the macro variable screen (refer to Sec. 1.3 for the method of displaying macro variables).

Though macro-programming features can be used in any program (the program given above is an example, which can be executed as the main program), a macro program or a subprogram is invariably designed to be called by other programs (main programs, subprograms, or macro programs). As mentioned in Sec. 3.3, and discussed in detail in Chaps. 6 and 7, a subprogram is called by M98, and a macro program (which is simply referred to as a macro) is called by G65 or G66. And, if a program is designed to be called by other programs, it must end with M99, rather than M30. So, **if any of the programs given in this chapter are to be**

called by other programs, replace M30 by M99 in the end. M30 has deliberately been used so that these programs could be run as main programs. M99 at the end of a main program runs the program repeatedly, in an infinite loop, until the RESET key is pressed.

Also recall that if a comment is inserted immediately after the O-word, as has been done in this program, the directory display (for directory display, press the PROG function key, followed by the DIR soft key, or press the PROG key twice, in the EDIT mode) on the MDI screen shows the inserted comment (up to 31 characters on 7.2-in LCD screen) along with the program number. This is a recommended practice, as one gets the information about the type of a program without opening it (i.e., O SRHing it).

The given program is quite general in nature. For example, if summation from 1 to 100 is desired, simply change the #3 value to 100 in the beginning of the program, and the obtained summation value would be 5050. A good macro program should provide for such flexibility in the very beginning of the program, so that other users (a macro is invariably designed to be used by several users), who may not be expert programmers, do not have to analyze the whole program to figure out where it needs to be edited.

As an exercise, the reader may try to write a program for calculating the sum of squares of the first 10 numbers (i.e., $1^2 + 2^2 + \cdots + 10^2$). The program would nearly be the same as the one given above for calculating the sum of the first 10 numbers. Only the #2 = #2 + #1 statement would need to be replaced by #2 = #2 + #1 * #1. The answer would be 385.

Though permanent common variables are very useful for applications such as the one just discussed, prestored values may not always be desirable, as this may result in incorrect initialization of variables in some cases (e.g., #500 = #500 + #1, where the initial value of #500 is desired to be 0). Moreover, it can never be said with certainty if the displayed values are actually calculated by the program or are just the old values. So, for the sake of safe programming practice, all the permanent common variables, which are to be used in a program, are set to null in the very beginning of the program, unless their old values are actually needed.

In some cases, it may even be desirable to clear all the permanent common variables. Program number 8001 does the same, setting all variables in the range #500 to #999 to null. The execution completes in about 10 seconds on 0i Mate TC (macro calculations are not very fast). The flowchart of an algorithm and its execution trace are shown in Fig. 5.2.

```
O8001 (CLEARING ALL PERMANENT COMM VARS);
(#1 @ COUNTER);
;
#1 = 500;                  (Initial value of COUNTER)
N10 #[#1] = #0;            (The variable number, referenced by
                            COUNTER, set to null)
```

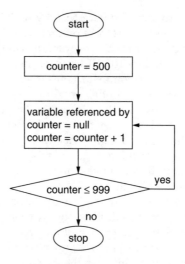

execution trace						
	counter	#500	#501	#502	⋯	#999
initial values of counter and variables	500	old	old	old	⋯	old
values after executing I loop	501	null	old	old	⋯	old
values after executing II loop	502	null	null	old	⋯	old
values after executing III loop	503	null	null	null	⋯	old
⋯	⋯	⋯	⋯	⋯	⋯	⋯
⋯	⋯	⋯	⋯	⋯	⋯	⋯
values after executing 499th loop	999	null	null	null	⋯	old
values after executing 500th loop	1000	null	null	null	⋯	null

FIGURE 5.2 Flowchart and execution trace of an algorithm for clearing all permanent common variables.

```
#1 = #1 + 1;
```
(COUNTER incremented by 1, for selecting the next variable)

```
IF [#1 LE 999] GOTO 10;
```
(Branches to N10 if #1 is less than or equal to 999, i.e., if all the variables up to #999 are not cleared)

```
M30;
```

A more complex example is to calculate the sample standard deviation (σ_{n-1}) of, say, 10 given numbers which are stored in #1 through #10. The mathematical formula is

$$\bar{x} = \frac{\sum x_i}{n}$$

$$\sigma_{n-1} = \sqrt{\frac{\sum(\bar{x} - x_i)^2}{n-1}}$$

where x_i = individual data

n = the total number of data

$\bar{x}$ = the average value of the given data

The flowchart of an algorithm for calculating σ_{n-1} is given in Fig. 5.3. Program number 8002 is based on this algorithm. Though any values can be stored in the variables #1 to #10, the chosen values in this example are 1, 2, 3, . . . , 10, respectively. The obtained answer would be 3.0276503. This program can be used for calculating σ_{n-1} for up to 33 numbers, stored in #1 through #33.

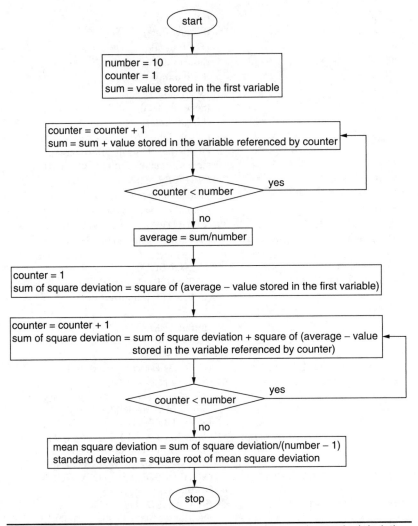

FIGURE 5.3 Flowchart of an algorithm for calculating the sample standard deviation.

```
O8002 (STANDARD DEVIATION CALCULATION);
(#1 THROUGH #10 @ DATA);
(STORE DESIRED VALUES IN #1...#10);
(#100 @ NUMBER OF GIVEN DATA);
(SET #100 TO THE NUMBER OF DATA);
(#101 @ COUNTER);
(#102 @ SUM OF GIVEN DATA);
(#103 @ AVERAGE OF GIVEN DATA);
(#104 @ SUM OF SQUARE DEVIATION);
(#105 @ MEAN SQUARE DEVIATION);
(#501 @ STANDARD DEVIATION);
;
```

#1 = 1;	(First data)
#2 = 2;	(Second data)
#3 = 3;	(Third data)
#4 = 4;	(Fourth data)
#5 = 5;	(Fifth data)
#6 = 6;	(Sixth data)
#7 = 7;	(Seventh data)
#8 = 8;	(Eighth data)
#9 = 9;	(Ninth data)
#10 = 10;	(Tenth data. Note that the data-entry step is not shown in the flowchart)
#100 = 10;	(NUMBER of data)
#101 = 1;	(Initial value of COUNTER)
#102 = #1;	(Initial SUM set to the first data)
N10 #101 = #101 + 1;	(COUNTER incremented by 1)
#102 = #102 + #[#101];	(Next data added to SUM)
IF [#101 LT #100] GOTO 10;	(Branching to N10 if all data are not added to SUM)
#103 = #102 / #100;	(AVERAGE calculated)
#101 = 1;	(COUNTER initialized to 1)
#104 = [#103 − #1] * [#103 − #1];	(Initial SUM OF SQUARE DEVIATION set to the square deviation for the first data)
N20 #101 = #101 + 1;	(COUNTER incremented by 1)
#104 = #104 + [#103 − #[#101]] * [#103 − #[#101]];	(Square deviation for the next data added to SUM OF SQUARE DEVIATION)
IF [#101 LT #100] GOTO 20;	(Branching to N20 if square deviations for all data are not added to SUM OF SQUARE DEVIATION)
#105 = #104 / [#100 − 1];	(MEAN SQUARE DEVIATION calculated)
#501 = SQRT[#105];	(STANDARD DEVIATION calculated and stored in the permanent common variable #501)
M30;	

Note that, on a 7.2-in LCD screen, it is not possible to continuously type more than 34 characters in one block, using the MDI keys. In a complex mathematical calculation, more than 34 characters in a block do appear. For example, in the standard deviation program, there is a block (the one next to the N20 block) with 40 characters. The only way to have such blocks in a program is to first type the program on a PC, and then download it to the CNC through the RS-232C port. This way, a block with practically any number of characters can be used in a program. If, however, the program is to be typed using MDI keys only, a long block (with more than 34 characters) would need to be split into smaller blocks. For example,

```
#104 = #104 + [#103 - #[#101]] * [#103 - #[#101]];
```

can be split into smaller blocks in the following manner, which will have the same effect:

```
#111 = #103 - #[#101];        (A new variable #111 introduced for
                               intermediate calculations)

#111 = #111 * #111;
#104 = #104 + #111;
```

The next example finds the roots of a quadratic equation in the form

$$ax^2 + bx + c = 0$$

The formula for the two roots is

$$x_1 = \frac{-b + \sqrt{b^2 - 4ac}}{2a} \quad \text{and} \quad x_2 = \frac{-b - \sqrt{b^2 - 4ac}}{2a}$$

The flowchart of an algorithm for finding the two roots is shown in Fig. 5.4. In program number 8003, which is based on this algorithm, any values can be assigned to the variables corresponding to a, b, and c. Here, 1, 1, and –2 have been taken, which give 1 and –2 as the solution for the two roots.

```
O8003 (QUADRATIC EQUATION SOLUTION);
(FORM @ A * X * X + B * X + C = 0);
(#1 @ A);
(#2 @ B);
(#3 @ C);
(ASSIGN DESIRED VALUES TO A,B,C);
(#100 @ DISCRIMINANT);
(#502 @ ROOT1);
(#503 @ ROOT2);
;
#1 = 1;                        (a, i.e., coefficient of x²)
```

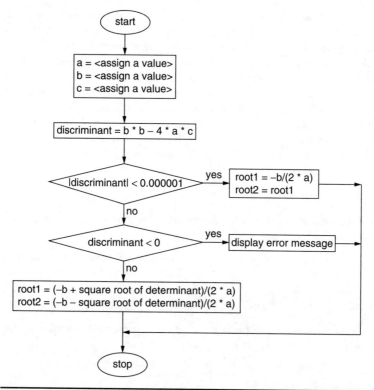

Figure 5.4   Flowchart of an algorithm for finding roots of a quadratic equation.

```
#2 = 1;                                (b, i.e., coefficient of x)
#3 = -2;                               (c, i.e., constant term)
#100 = #2 * #2 - 4 * #1 * #3;
                                       (Determinant calculated)
IF [ABS[#100] LT 0.000001] GOTO 10;
                                       (Branching to N10 if determinant
                                       is 0)
IF [#100 LT 0] THEN #3000 = 1 (IMAG ROOTS);
                                       ("3001 IMAG ROOTS" displayed
                                       and execution terminated, for a
                                       negative determinant. IF_THEN_
                                       statement is described in detail in
                                       Sec. 5.3)
#502 = [-#2 + SQRT[#100]] / [2 * #1];
                                       (First root calculated and stored
                                       in the permanent common vari-
                                       able #502)
```

```
#503 = [-#2 - SQRT[#100]] / [2 * #1];
```
(Second root calculated and stored in another permanent common variable #503)

```
GOTO 20;
```
(Branching to N20)

```
N10 #502 = -#2 / [2 * #1];
```
(First root calculated, when the determinant is 0)

```
#503 = #502;
```
(The two roots are equal)

```
N20 M30;
```

Recall that every macro calculation typically involves an error of 10^{-8}. So, even if the determinant is actually 0, it may not come out to be **exactly** 0. It is because of this reason that the absolute value of the determinant is being checked against a small value, in this program.

Note that, even if the determinant is negative, generating an alarm, the permanent common variables #502 and #503 would still have old values (if any) stored in them. If this is likely to cause a confusion, set #502 and #503 to null (#0), if the determinant comes out to be negative. In fact, as already stated, a good programming practice would be to **unconditionally** set these variables to null in the very beginning of the program. This applies to all programs which use permanent common variables, without needing their old values.

For additional practice, the reader may try to develop programs for finding

- The largest and the smallest number in a given set of data
- The sum of a finite series of the type $\sum_1^N$ (an expression in terms of index i)
- The sum of an infinite but convergent series (the calculation would need to be terminated when a series term is found to be very small, say, less than 0.000001)
- The calculation of factorial of a given number, etc.

Wherever possible, *recurrence relation* should be used for computational efficiency. For example, the factorial of a given number can be defined in terms of the factorial of its preceding number: $i! = (i-1)! \times i$. Such a technique is especially advantageous in series summation. An example is given below:

$$\text{sum} = x - \frac{x^3}{3!} + \frac{x^5}{5!} - \frac{x^7}{7!} + \cdots (-1)^{n-1} \frac{x^{2n-1}}{(2n-1)!}$$

An inspection indicates that

$$\text{ith term} = (-1)^{i-1} \frac{x^{2i-1}}{(2i-1)!} \quad \text{and} \quad (i+1)\text{th term} = (-1)^i \frac{x^{2i+1}}{(2i+1)!}$$

So, the (i + 1)th term can be expressed in terms of ith term in the following manner:

$$(i + 1)\text{th term} = (-1)\frac{x^2}{(2i+1)2i} \times \text{ith term}$$

The flowchart based on this recurrence relation is given in Fig. 5.5. The input data are given values for x and the number of terms (n) in the series. The flowchart is valid for $n \geq 2$. It is left as an exercise for the reader to write a program for finding the sum of this series.

Several of the given and the suggested programs are for the sole purpose of explaining the use of conditional statements in a macro, apart from describing the proper methodology for developing a macro program (in practice, no CNC machine may ever require the use of some of these programs). These programs were selected mainly because every engineering graduate writes such programs in some programming course. And, when the programs are familiar, one only has to see how these can be converted into the macro language. This makes learning a new language easy.

These programs also indicate that the macro-programming language can be used for doing complex calculations. In fact, the discussion that follows will show that a macro program can do much more than what is described here. So far, we have had only an introduction to this language.

FIGURE 5.5
Flowchart of an algorithm for the sum of a series.

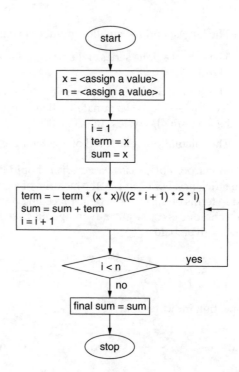

5.3   Conditional Execution of a Single Macro Statement

Such a statement (IF_THEN_) has already been used in program number 8003 (it is left as an exercise for the reader to think how the program can be developed without using this statement). Here, it is formally described, in detail.

A specified macro statement may or may not be executed depending on whether a specified condition is satisfied or not. The format is

```
IF [<a conditional expression>] THEN <a macro statement>;
```

The conditional expression may even be a complex combination of several independent conditions.

The IF_THEN_ format has the following limitations:

- Only a **single** macro statement can be specified.
- An NC statement is **not** allowed.

Examples:

```
IF [#1 LT #2] THEN #3 = 3;
```
(If the condition is TRUE, a value of 3 is assigned to the variable #3. A FALSE condition leaves the previous value stored in #3 unchanged)

```
IF [#1 LT #2] #3 = 3;
```
(Illegal command. THEN must be there)

```
IF [#1 LT #2] THEN M30;
```
(Illegal command because an NC statement has been specified)

```
IF [#1 LT #2] THEN [#3 = 3; #4 = 4];
```
(Illegal command because more than one macro statements have been specified. In fact, it is not possible to have this command displayed in the manner shown. The first semicolon changes the line which breaks the statement into two separate blocks, making it completely meaningless)

If more than one macro statements and/or NC statements are desired to be executed if a specified condition is TRUE (e.g., if [#1 LT #2] is TRUE), this can be done in the following manner (where the statements between the IF_GOTO_ and N100 blocks are executed only if #1 is less than #2):

```
IF [#1 GE #2] GOTO 100;
<macro statement -1>
<macro statement -2>
. . .
. . .
<macro statement -n>
<NC statement -1>
<NC statement -2>
. . .
. . .
```

```
<NC statement n>
N100;
<remaining part of the program>
```

Recall that an isolated EOB symbol (;) in a block, with or without a sequence number (such as N100; or simply ;), is treated as a blank **NC statement** which does not do anything, but has all the properties of an NC statement. For example, in the single-block mode, the execution would halt here, and wait for the CYCLE START button to be pressed again for resuming the execution. This can also be used to prevent prereading of macro statements, which may not be desirable in certain cases. This issue is discussed in more detail in Sec. 7.6.

Since GOTO n is a macro statement, it is also permissible to command

```
IF [<a conditional expression>] THEN GOTO n;
```

which is equivalent to

```
IF [<a conditional expression>] GOTO n;
```

5.4   Execution in a Loop

It is possible to execute certain lines of a program repeatedly (i.e., in a loop) so long as a specified condition remains satisfied. This has, in fact, already been done in several of the programs discussed so far. The conditional branching, IF_GOTO_, was used for this purpose. However, the use of GOTO (especially unconditional GOTO) is generally not recommended as it makes the program less readable, making its interpretation a bit difficult. A better way to construct a loop is to use the WHILE statement. And, this is the only alternate way available in Custom Macro B, for executing in a loop.

Another advantage of creating a loop using the WHILE statement is that its processing takes less time compared to the processing of the GOTO statement. This might affect the speed of program execution if the number of times the loop is required to be executed is too large. The syntax of the WHILE statement is

```
WHILE [<a conditional expression>] DO n;
...
...
<program blocks within the loop>
...
...
END n;
```

where n can be 1, 2, or 3.

If the specified condition is satisfied (i.e., if the conditional expression evaluates to TRUE), the program blocks between DO and END are executed. Thereafter, the condition is checked again, and if the condition is still satisfied, the loop is reexecuted. As long as the condition remains satisfied (which is tested in the **beginning** of each loop), the loop is repeatedly executed.

The program blocks inside the loop are so designed that their processing affects the specified condition, such that, after a few executions of the loop, the condition becomes FALSE. Thereafter, the program execution proceeds to the block next to END, after **fully** completing the execution of the loop which changed the condition from TRUE to FALSE. Note that the condition is checked before executing a new loop; it is not checked while executing a loop. The flowchart for the WHILE statement is given in Fig. 5.6.

Though only 1, 2, and 3 can be used as identification numbers (n), these can be used any number of times in a program:

```
WHILE [<conditional expression 1>] DO 1;
```
(The identification number is 1)
```
...
...
<program blocks within the first loop>
...
...
END 1;
WHILE [<conditional expression 2>] DO 1;
```
(Same identification number permitted)
```
...
...
```

FIGURE 5.6
Flowchart for a
WHILE statement.

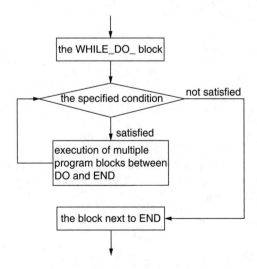

```
<program blocks within the second loop>
...
...
END 1;
```

In fact, the same identification number (say, 1) can be used every-where in the program (as done in the previous example), except in *nesting* where 1, 2, and 3 correspond to the three levels of nested WHILE statements (up to a maximum of three levels of nesting is permitted):

```
WHILE [<condition 1> DO 1;
...
...
WHILE [<condition 2>] DO 2;
...
...
WHILE [<condition 3>] DO 3;
...
...
END 3;
...
...
END 2;
...
...
END 1;
```

level 3 level 2 level 1

The DO_END_ ranges cannot overlap in nesting, as it makes the statement illogical and meaningless:

```
WHILE [<condition 1>] DO 1;
...
...
WHILE [<condition 2>] DO 2;
...
...
END 1;
...
...
END 2;
```

overlapping ranges, which makes the nesting illegal

It is permitted to use any macro or NC statement, including IF_ THEN_, IF_GOTO_ and GOTO_, inside a loop (i.e., inside the DO_ END_ range):

```
WHILE [<condition 1>] DO 1;
...
...
```

```
IF [<condition 2>] GOTO 10;      (Permitted to use GOTO inside a
                                  loop)
. . .
. . .
N10 ...;
. . .
. . .
END 1;
```

It is also permitted to jump out of the loop using a GOTO statement:

```
WHILE [<condition 1>] DO 1;
. . .
. . .
IF [<condition 2>] GOTO 10;      (Permitted to jump outside the
                                  loop)
. . .
. . .
END 1;
. . .
. . .
N10 ...;
```

However, jumping inside a loop (from outside) is not permitted:

```
IF [<condition 1>] GOTO 10;      (Illegal jump inside a loop)
. . .
. . .
WHILE [<condition 2>] DO 1;
. . .
. . .
N10 ...;
. . .
. . .
END 1;
```

For illustrating the use of the WHILE statement, the same problems would be considered to bring out the difference from the loops created using the conditional GOTO statement.

As already stated, the program coding should be done only after preparing a flowchart. The flowchart for the WHILE statement would be slightly different because the specified condition is checked in the **beginning** of the loop.

The flowchart for the summation problem $(1 + 2 + 3 + \cdots + 10)$ is given in Fig. 5.7. Program number 8004 is based on this flowchart.

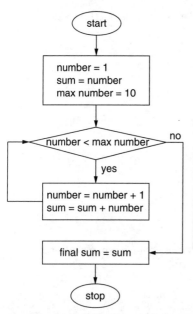

FIGURE 5.7 Flowchart and execution trace of a summation algorithm for a WHILE statement.

execution trace		number	sum
initial values of number and sum		1	1
values after executing I loop		2	3
values after executing II loop		3	6
values after executing III loop		4	10
values after executing IV loop		5	15
values after executing V loop		6	21
values after executing VI loop		7	28
values after executing VII loop		8	36
values after executing VIII loop		9	45
values after executing IX loop		10	55
final sum			55

```
O8004 (SUMMATION OF NUMBERS USING WHILE);
(#1 @ NUMBER);
(#2 @ SUM);
(#3 @ MAX NUMBER);            (Up to which summation is desired)
(#504 @ FINAL SUM);
;
#1 = 1;                       (Initial value of NUMBER)
#2 = #1;                      (Initial value of SUM)
#3 = 10;                      (Summation from 1 to 10 is desired)
WHILE [#1 LT #3] DO 1;        (The loop starts here. It is executed if #1
                              is less than #3. If not, the loop termi-
                              nates, and the execution jumps to the
                              block after END 1)
#1 = #1 + 1;                  (NUMBER incremented by 1)
#2 = #2 + #1;                 (SUM calculated)
END 1;                        (The end of the loop. The execution
                              jumps to the start of the loop to check if
                              #1 is still less than #3)

#504 = #2;                    (The final value of the SUM stored in
                              the permanent common variable #504)
M30;
```

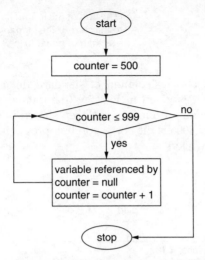

execution trace						
	counter	#500	#501	#502	⋯	#999
initial values of counter and variables	500	old	old	old	⋯	old
values after executing I loop	501	null	old	old	⋯	old
values after executing II loop	502	null	null	old	⋯	old
values after executing III loop	503	null	null	null	⋯	old
⋯	⋯	⋯	⋯	⋯	⋯	⋯
⋯	⋯	⋯	⋯	⋯	⋯	⋯
values after executing 499th loop	999	null	null	null	⋯	old
values after executing 500th loop	1000	null	null	null	⋯	null

Figure 5.8 Flowchart and execution trace of an algorithm for clearing all permanent common variables using a WHILE statement.

The flowchart for clearing all permanent common variables using a WHILE statement is given in Fig. 5.8. Program number 8005 is based on this flowchart. The execution of this program takes about 5 seconds which is nearly half of the time taken by program number 8001 which used a GOTO statement to construct the loop.

```
O8005 (CLEARING COMM VARS USING WHILE);
(#1 @ COUNTER);
;
```

`#1 = 500;`	(Initial value of COUNTER)
`WHILE [#1 LE 999] DO#1;`	(The loop starts here. It is executed if #1 is less than or equal to 999. If not, the loop terminates, and the execution jumps to the block after END 1)
`#[#1] = #0;`	(The variable number, referenced by COUNTER, cleared to null)
`#1 = #1 + 1;`	(COUNTER incremented by 1, for selecting the next variable)

```
END#1 ;
```
(The end of the loop. The execution jumps to the start of the loop to check if #1 is still less than or equal to 999)

```
M30 ;
```

The flowchart for calculation of standard deviation using the WHILE statement is given in Fig. 5.9 (the data-entry step, in the beginning of the algorithm, has not been shown in the flowchart). Program number 8006 is the corresponding program. Note that this algorithm uses two loops.

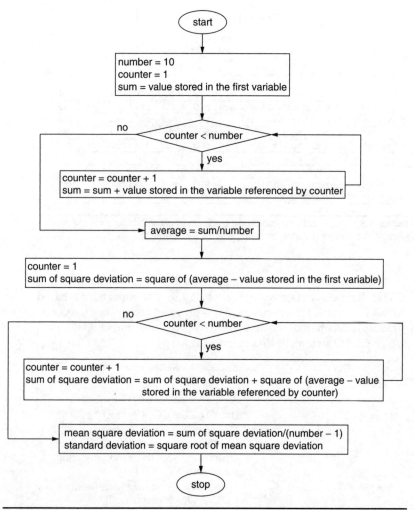

FIGURE 5.9 Flowchart of an algorithm for calculating the sample standard deviation using a WHILE statement.

```
O8006 (STD DEV CALCULATION USING WHILE);
(#1 THROUGH #10 @ DATA);
(STORE DESIRED VALUES IN #1...#10);
(#100 @ NUMBER OF GIVEN DATA);
(SET #100 TO THE NUMBER OF DATA);
(#101 @ COUNTER);
(#102 @ SUM OF GIVEN DATA);
(#103 @ AVERAGE OF GIVEN DATA);
(#104 @ SUM OF SQUARE DEVIATION);
(#105 @ MEAN SQUARE DEVIATION);
(#505 @ STANDARD DEVIATION);
;
```

`#1 = 1;`	(First data)
`#2 = 2;`	(Second data)
`#3 = 3;`	(Third data)
`#4 = 4;`	(Fourth data)
`#5 = 5;`	(Fifth data)
`#6 = 6;`	(Sixth data)
`#7 = 7;`	(Seventh data)
`#8 = 8;`	(Eighth data)
`#9 = 9;`	(Ninth data)
`#10 = 10;`	(Tenth data)
`#100 = 10;`	(NUMBER of data)
`#101 = 1;`	(Initial value of COUNTER)
`#102 = #1;`	(Initial SUM set to the first data)
`WHILE [#101 LT #100] DO 1;`	(The first loop starts here. It ends at the END 1 block. Note that END 1 appears twice in this program because the same identification number, 1, has been used for both the loops. The current loop is executed if #101 is less than #100. If not, the loop terminates, and the execution jumps after the END 1 block of this loop)
`#101 = #101 + 1;`	(COUNTER incremented by 1)
`#102 = #102 + #[#101];`	(Next data added to SUM)
`END 1;`	(The end of the first loop. The execution jumps to the start of the loop to check if #101 is still less than #100)
`#103 = #102 / #100;`	(AVERAGE calculated)
`#101 = 1;`	(COUNTER initialized to 1)
`#104 = [#103 − #1] * [#103 − #1];`	(Initial SUM OF SQUARE DEVIATION set to the square deviation for the first data)
`WHILE [#101 LT #100] DO 1;`	(The second loop starts here. It is executed if #101 is less than #100. If not, the loop terminates, and the execution jumps to the block after the END 1 block of this loop, that is,

jumps after the second END 1 of this program)

```
#101 = #101 + 1;                    (COUNTER incremented by 1)
#104 = #104 + [#103 - #[#101]] * [#103 - #[#101]];
```
(Square deviation for the next data added to SUM OF SQUARE DEVIATION)

```
END 1;
```
(The end of the second loop. The execution jumps to the start of the loop to check if #101 is still less than #100)

```
#105 = #104 / [#100 - 1];
```
(MEAN SQUARE DEVIATION calculated)

```
#505 = SQRT[#105];
```
(STANDARD DEVIATION calculated and stored in the permanent common variable #505)

```
M30;
```

The series summation flowchart shown in Fig. 5.5 was designed to be used with the conditional GOTO statement. The same algorithm, if desired to be implemented with the WHILE statement, would have a flowchart as shown in Fig. 5.10. The reader should try to write a program, based on this flowchart, which is being left as an

---

Figure 5.10
Flowchart of an algorithm for the sum of a series using a WHILE statement.

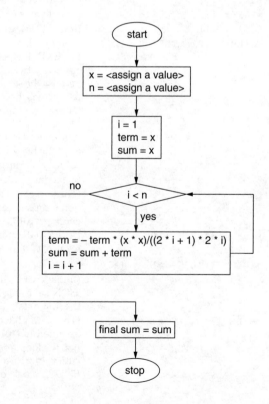

exercise. Just define variables for inputs x and n, as well as for counter i, term, and sum. Thereafter, program writing is simply a matter of
copying the flowchart in terms of the defined variables.

Though the WHILE statement is for the purpose of executing in a
loop, it can also be used to execute certain blocks **once** if a specified
condition is true:

```
WHILE [<a conditional expression>] DO 1;
...
...
...
GOTO 10;
END 1;
N10 ...;
```

Recall that the IF_THEN_ format can execute **only one** macro statement if a specified condition is satisfied. An indirect method was suggested for executing more than one statement.

5.5   Arithmetic Operations on Macro Variable Numbers

As already discussed, macro variables cannot be defined using arbitrary combinations of alphanumeric characters. These can only be
defined in terms of certain variable numbers. This is a bit inconvenient
because the programmer does not get any intuitive feeling of the
assigned meanings of the different variables being used in a program.

However, a unique feature of macro variables is that arithmetic
operations are permitted on the variable numbers. This can be used
advantageously in certain cases. For example, in PASCAL, one may
store the marks in five subjects obtained by a student in five variables, say, MARKS1, MARKS2, MARKS3, MARKS4, and MARKS5.
On the other hand, in a macro program, the marks may be stored in,
say, #101, #102, #103, #104, and #105. While MARKS1, MARKS2,
MARKS3, MARKS4, and MARKS5 are all independent of one another,
#101, #102, #103, #104, and #105 follow a pattern. The variable num-
ber for a subject can be obtained by adding 1 to the variable number
for the previous subject. This fact can be used while writing a program for, say, calculating the aggregate marks. The flowchart of such
an algorithm is given in Fig. 5.11, and the corresponding program
number is 8007. The WHILE statement has been used for creating the
loop in this program, as it is preferred over the conditional GOTO
statement. The readers should, however, try to construct the loop
using the IF_GOTO_ statement, as an exercise.

```
O8007 (AGGREGATE MARKS CALCULATION);
(#1 @ COUNTER);
(#2 @ SUM);
```

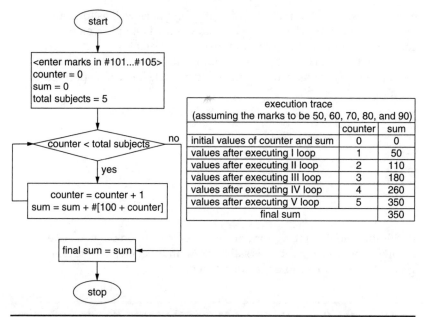

Figure 5.11 Flowchart and execution trace of an algorithm for calculating aggregate marks.

```
(#3 @ TOTAL SUBJECTS);
(#506 @ FINAL SUM);
;
#101 = 50;
#102 = 60;
#103 = 70;
#104 = 80;
#105 = 90;
#1 = 0;
#2 = 0;
#3 = 5;                       (If the total number of subjects is differ-
                             ent from 5, define #3 accordingly)

;
WHILE [#1 LT #3] DO 1;        (Start of the loop)
#1 = #1 + 1;                 (COUNTER incremented by 1)
#2 = #2 + #[100 + #1];       (Marks obtained in the next subject
                             added to SUM)

END 1;                       (End of the loop)
;
#506 = #2;                   (FINAL SUM stored in the permanent
                             common variable #506)

M30;
```

This, of course, is a trivial example; the marks in the five subjects can be added straightaway! However, if the number of data is too large (say, 100), such a program would be useful. Moreover, the objective here is to explain how arithmetic operations are done on variables numbers, and how the flowchart, involving a loop, in such a case is drawn.

Such a feature of macro variables can also be used for manipulating certain system variables which follow a definite pattern. For example, there is a jump of 20 in the variable numbers for various work offset values for a particular axis (see each row of Table 5.2(a), which is a part of Table 3.12). Similarly, corresponding to some specific workpiece coordinate system, the variable number increments by 1 for the next axis (see each column of Table 5.2(a). The identified pattern can be described mathematically, in terms of number 5201, as shown in Table 5.2(b), which is a part of Table 5.2(a). So, if the axis and the workpiece coordinate system are specified, the corresponding system variable number can be calculated. The multiplication factors (of 20) for external G54, G55, G56, G57, G58, and G59 workpiece coordinate systems are 0, 1, 2, 3, 4, 5, and 6, respectively, and the addends are 0, 1, and 2 for X-, Y-, and Z-axis, respectively (3, 4, 5, etc. for the remaining axes, if available). So, for example, the Y-axis offset value for the G56 coordinate system is stored in variable number $5201 + 20 \times 3 + 1$, i.e., in #5262.

Axis	External	G54	G55	G56	G57	G58	G59
X	#5201	#5221	#5241	#5261	#5281	#5301	#5321
Y	#5202	#5222	#5242	#5262	#5282	#5302	#5322
Z	#5203	#5223	#5243	#5263	#5283	#5303	#5323

(a)

Axis	External	G54	G55	G56
X	#[5201 + 20 × 0 + 0]	#[5201 + 20 × 1 + 0]	#[5201 + 20 × 2 + 0]	#[5201 + 20 × 3 + 0]
Y	#[5201 + 20 × 0 + 1]	#[5201 + 20 × 1 + 1]	#[5201 + 20 × 2 + 1]	#[5201 + 20 × 3 + 1]
Z	#[5201 + 20 × 0 + 2]	#[5201 + 20 × 1 + 2]	#[5201 + 20 × 2 + 2]	#[5201 + 20 × 3 + 2]

(b)

TABLE **5.2** Pattern in System Variable Numbers for Various Work Offset Values on a Milling Machine

Methods of Zero Shift

We will now consider a simple example to see how such system variables can be identified and manipulated for a desired effect through a program. Let us say, the requirement is to place the origin of the currently active (or some other) workpiece coordinate system (among G54–G59 and optionally available 48 additional coordinate systems G54.1 P1–P48) at the current XY-position of the tool, on a milling machine. This is known as *zero shift* or *datum shift*. Recall that the origin of the workpiece coordinate system (G54, G55, etc.) is also referred to as the *workpiece zero point* or the *program zero point* or the *component zero point*, and at any time, any one of the six standard and 48 additional workpiece coordinate systems can be selected to be the active coordinate system. A program is executed in the currently active coordinate system (CNC interprets the coordinate values used in a program in the currently active coordinate system).

Zero shift can be very simply done just by commanding G90 G92 X0 Y0, which is obviously the simplest method. The corresponding command on a lathe is G50 X0 Z0. Zero shift for single axis is also possible. For example, G90 G92 Z0 (G50 Z0 on a lathe) shifts zero only for the Z-axis. However, a limitation of using G92 or G50 is that the shift remains valid only for the **current machining session**. Once the machine is switched off and again switched on, the information fed to it through G92 or G50 in the previous session is lost, and the machine starts using the original zero setting. Moreover, zero shift, using this method, can only be done for the **currently active coordinate system**.

On the other hand, if zero shift is done by manipulating the values stored in the system variables for the relevant work offset distances, the shifted origin remains fixed at the new position **permanently**, until the offset values are changed again (using any method). In addition, this method works for all coordinate systems, **even if these are not currently active**. Of course, if the system variables, corresponding to the offset distances of a coordinate system which is not currently active, are manipulated, its effect will be seen only after this coordinate system becomes active.

The system variables for various offset values are read/write type. So, if the values stored in them are changed, the corresponding offset distances automatically change. Conversely, if the offset distances are changed, the corresponding system variables automatically get redefined (i.e., store new values). Note that the terms "offset distance," "offset value," and even "offset" are usually used synonymously. So, X-offset value, X-offset distance, and X-offset all mean the same thing.

Further discussion on this topic necessitates a clear understanding of the way the control defines and uses the various coordinate systems. Refer to Fig. 5.12 for the discussion that follows. The X- and Y-offset distances are the X- and Y-component (with sign) of the corresponding offset vector shown in this figure. For a lathe, replace X by Z, and Y by X.

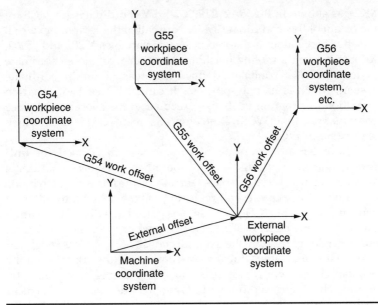

FIGURE 5.12 Coordinate systems and offsets on a milling machine.

Machine Coordinate System

On any machine, the origin of the *machine coordinate system* (MCS), which is also called the *machine zero point*, is placed at a desired location by a parameter setting. This is done by the MTB, and there is no need for the end-user to change it. In fact, it must not be changed because several machine settings are done in MCS.

For example, the 1300 series parameters (refer to the *Parameter Manual* for details) are used for defining the software overtravel limits for the tool movement. If, due to a programming error or otherwise, the tool is required to enter the prohibited zone, it will automatically stop at the boundary, with an alarm. Such a software limit is desirable because of safety considerations. For example, the tool should never hit the chuck or the tailstock on a lathe. The software limits do not allow the tool to enter a defined dangerous zone. These limits are specified in certain parameters, as coordinate values in MCS. This means that any change in the position of the MCS would automatically redefine the prohibited zone, which is obviously not desirable. So, one can say that **the MCS is fixed for a given machine**.

The MCS, however, is not convenient for running part programs; in fact, the control does not allow program execution in MCS. There are other coordinate systems for this purpose.

External Workpiece Coordinate System

The *external workpiece coordinate system*, which is usually referred to simply as the *external coordinate system*, is defined relative to

MCS, as shown in Fig. 5.12. The X- and Y-coordinates (in MCS) of the origin of this coordinate system are called the X- and Y-external offsets. These values are stored in system variables #5201 and #5202, respectively, on a milling machine. All other workpiece coordinate systems (G54, G55, etc.) are defined with respect to this coordinate system, which is its only purpose. It cannot be used for program execution. A program is always executed in the chosen workpiece coordinate system (WCS). If no choice is made, G54 is used which is the default WCS.

The external workpiece coordinate system is used for **uniformly** shifting the origins of all the other workpiece coordinate systems. Such a situation arises when the same fixture, which holds a number of workpieces, all requiring to be machined in different coordinate systems (G54, G55, etc.), is placed at a different location on the machine table. In such cases, instead of redefining G54, G55, etc. independently, only the external coordinate system is redefined to have the same effect. Such a situation on a lathe arises when the original chuck, which was used for offset setting, is replaced by another chuck with a different holding position along the Z-axis. If such a requirement is not there, there is no need to use this coordinate system. In such cases, all the external offset values are kept zero, which means that this coordinate system coincides with the MCS, and G54, G55, etc. are, effectively, directly defined with respect to the MCS.

G54–G59, G54.1 P1–P48 Workpiece Coordinate Systems

Six standard and 48 optional (which are activated, on milling machines only, on extra payment) workpiece coordinate systems are available. These are all defined with respect to the external workpiece coordinate system, as shown in Fig. 5.12. The various offset distances are stored in different system variables, as shown in Tables 3.12 and 5.2 (which is a subset of Table 3.12).

These coordinate systems are used for machining purpose (only), and placed at desired locations (i.e., at the desired datum of the workpiece) by specifying the various offset distances. The procedure of tool offset setting actually measures these offset distances only.

Only one of these coordinate systems remains active at a time. The coordinates used in the program are interpreted by the control, in the active coordinate system (say, in the G54 coordinate system, though the control internally converts all coordinates in the machine coordinate system). If there is only a single workpiece to be machined all the time, there is no need to have so many workpiece coordinate systems. However, in a practical situation, several workpieces need to be machined, all having data at different locations. In such cases, independent coordinate systems, corresponding to every datum, are defined, and the appropriate coordinate system is made active whenever required. Since G54 is the default coordinate system, whenever the machine is switched on, or the control is the reset (by pressing RESET key, or by executing M02 or M30), G54 becomes active.

Zero Shift by Manipulating Offset Distances

By now, it should be clear that the control places the origins of different WCSs at the **user-specified** distances from the origin of the MCS. This distance is the vector sum of the external offset vector and the corresponding work offset vector. So, for example, if the X-external offset is 10, and the X-offset for G54 is 200, then the control will place the origin of the G54 coordinate system at a distance of 210, in the positive X-direction, from the origin of the MCS. On the other hand, if the desired distance of G54 from the MCS is, say, 250, and the external offset is not to be changed, then the X-offset of G54 will have to be changed to 240. Therefore, as an example of zero shift, if the current X-position of the tool is to be made the origin of G54 (which may or may not be the currently active coordinate system), the **sum** of the X-external offset and the X-offset of G54 will have to be made equal to the X-distance of the current position of the tool from the MCS (which will be equal to the X-coordinate of the current tool position in MCS). Mathematically,

X-external offset + X-offset of G54 = X-coordinate of the
current tool position, in MCS

Hence, if the external offset value is not to be changed, the X-offset of G54 will have to be made equal to the X-coordinate in MCS minus the X-external offset, that is,

X-offset of G54 = X-coordinate of the current tool position,
in MCS – X-external offset

Similarly, for zero shift to the current Y-position of the tool,

Y-offset of G54 = Y-coordinate of the current tool position,
in MCS – Y-external offset

This is the principle we are going to make use of for zero shift, that is, for shifting the origin of the desired (active or otherwise) WCS to the current tool position.

Zero Shift through System Variables

For placing the origin of the current WCS (though the method is valid also for a coordinate system which is not currently active) at the current tool position, let us assume, for the sake of illustration, that the currently active coordinate system is G56, and the current XY-position of the tool is to be made the new datum on a milling machine. This can be done by redefining X- and Y-offset values of G56, using the following formulae:

X-offset of G56 = X-coordinate of the current tool position,
in MCS – X-external offset

Y-offset of G56 = Y-coordinate of the current tool position,
in MCS – Y-external offset

In terms of system variables, these translate to (see Table 3.11 and Table 3.12/Table 5.2)

```
#5261 = #5021 - #5201;
#5262 = #5022 - #5202;
```

Note that the RHS of the formulae given above would be the same for all the WCSs. For example, these formulae will take the following form for XY-datum shift for G59:

```
#5321 = #5021 - #5201;
#5322 = #5022 - #5202;
```

Work Offset Display on a Milling Machine

A typical work offset screen on a milling machine, when the active coordinate system is G56, is shown in Table 5.3. The components of the external offset vector and the various work offset vectors, along different axes, are the corresponding offset values, which are displayed on this screen. The X-, Y-, and Z-offset values corresponding

WORK COORDINATES					
(G56)					
NO.		DATA	NO.		DATA
	X	0.000		X	
00	Y	0.000	02	Y	
(EXT)	Z	0.000	(G55)	Z	
	X			X	
01	Y		03	Y	
(G54)	Z		(G56)	Z	

TABLE 5.3   A Typical Work Offset Screen on a Milling Machine

to G54, G55, and G56 have been left blank in this table. These will actually display some values. The external offset values will be all zero, as displayed, if the external coordinate system is not being used, which is usually the case. As already explained, the purpose of using the external coordinate system is to shift the datum of **all** the other workpiece coordinate systems, G54 to G59 as well as G54.1 P1–P48, by the **same** specified amount.

The work offset screen can be displayed by pressing the OFS/SET key, followed by the WORK soft key. Note the display of G56 at the top left corner of this screen, which indicates that G56 is currently active.

System Variables for Various Work Offset Values

The system variable numbers, corresponding to various offset values, are shown in Table 5.4, which is a deliberate repetition of Table 5.2(*a*), in a different form, for the sake of better clarity, as one can directly correlate the system variables with what one sees on the screen. The variable numbers corresponding to G57, G58, and G59, on a milling machine, are shown in Table 5.5.

Note that the variable numbers corresponding to various offset values do not change when the current coordinate system changes. So, for example, #5261 always contains the X-axis offset value for G56, irrespective of which coordinate system is currently active. Of course, the machine uses the value stored in #5261 only when G56 becomes the current coordinate system.

Zero Shift through a Program

For zero shift, the offset values corresponding to the currently active workpiece coordinate system will need to be edited (editing is possible for any WCS, but we are considering the current WCS in our

WORK COORDINATES					
(G56)					
NO.		DATA	NO.		DATA
	X	#5201		X	#5241
00	Y	#5202	02	Y	#5242
(EXT)	Z	#5203	(G55)	Z	#5243
	X	#5221		X	#5261
01	Y	#5222	03	Y	#5262
(G54)	Z	#5223	(G56)	Z	#5263

TABLE **5.4** System Variables Corresponding to External, G54, G55, and G56 Work Offset Values on a Milling Machine

WORK COORDINATES					
(G56)					
NO.		**DATA**	**NO.**		**DATA**
	X	#5281		X	#5321
04	Y	#5282	06	Y	#5322
(G57)	Z	#5283	(G59)	Z	#5323
	X	#5301			
05	Y	#5302			
(G58)	Z	#5303			

TABLE 5.5 System Variables Corresponding to G57, G58, and G59 Work Offset Values on a Milling Machine

example). This can be done manually also, after highlighting it by arrow keys, by typing the new value, and pressing the INPUT soft key or the INPUT key on the MDI panel. When this is desired to be done through a program, one needs to have the following information, inside the program:

- The currently active workpiece coordinate system
- The system variable numbers, corresponding to the work offset values of this coordinate system
- The coordinates of the current tool position in MCS

Referring to Table 3.10(b), the system variable #4014 stores the number of the current workpiece coordinate system (which would be 56 in our example, since G56 is assumed to be the current coordinate system). On the other hand, the system variables #5021 and #5022 store the current X- and Y-coordinates (on a milling machine), respectively, in MCS (see Table 3.11). Program number 8008, which is based on this information, shifts the origin of the currently active coordinate system to the current tool position on a milling machine.

```
O8008 (CURR WCS DATUM SHIFT ON MILL M/C);
    #1 = #4014;              (The current coordinate system, in
                             our example, is G56. So, 56 would
                             get stored in #1)

    #1 = #1 - 53;           (This would store 1 in #1, if the cur-
                             rent coordinate system is G54. Simi-
                             larly, corresponding to G55, G56,
                             G57, G58, and G59, the stored val-
                             ues would be 2, 3, 4, 5, and 6, respec-
                             tively. In our example, #1 would
                             store 3)
```

```
#1 = #1 * 20;
```

(This calculates the multiplication factor, as defined in Table 5.2(b), for the current coordinate system. In our example, #1 would store 60)

```
#1 = #1 + 5201;
```

(#1 would store 5221, 5241, 5261, 5281, 5301, and 5321, corresponding to G54, G55, G56, G57, G58, and G59, respectively, depending on which one is the current coordinate system. The stored value would be the variable number corresponding to the X-axis work offset value for the current coordinate system, as shown in Table 3.12/Table 5.2(a). In our example, #1 would store 5261)

```
#[#1] = #5021 - #5201;
```

(#5021 contains the tool's current X-coordinate in MCS, and #5201 contains the X-external offset. As already explained, when the difference in their values is stored in the system variable corresponding to the X-offset of a WCS, the current position of the tool becomes the new X-datum of **that WCS**, that is, the current tool position gets redefined as X0 in that WCS. In our example, #5261 would get appropriately modified, which would shift the X-datum for G56 to the current X-position of the tool)

```
#[#1 + 1] = #5022 - #5202;
```

(Shifts datum for the Y-axis to the current Y-position of the tool. In our example, #5262 would get appropriately modified, which would shift the Y-datum of G56)

```
M30;
```

Note that this program has been written without making a flowchart. A flowchart is not needed for straightforward cases such as this. The algorithm in this program is very simple:

1. Identify the current workpiece coordinate system.

2. Identify the system variables for X- and Y-work offset values corresponding to this coordinate system.

3. Redefine the identified system variables, using the given formulae.

However, even if one decides to write a program without making a flowchart, at least the algorithm must be noted down in black and white, before attempting to write the program. This would be useful for future reference also, as a quick intuitive interpretation of a macro program is generally not possible due to its abstract look.

Though the program given above is meant to be used on a milling machine, it would work on a lathe in pretty much the same manner. The only difference is that the second axis on a lathe is the Z-axis. So, the program would be exactly same. Just replace Y by Z in the description for this program. However, in a practical application on a lathe, normally only Z-datum shift might be needed. If so, delete #[#1] = #5021 − #5201 block of the program. Program number 8009 is the modified program.

```
O8009 (CURR WCS Z-DATUM SHIFT ON LATHE);
#1 = #4014;
#1 = #1 - 53;
#1 = #1 * 20;
#1 = #1 + 5201;
#[#1 + 1] = #5022 - #5202;
M30;
```

Both the programs are designed to identify and shift the datum of the **current** WCS. If the requirement is to shift the datum of, say, G56 coordinate system, irrespective of which WCS is current, this is a simpler task, which can be done by program numbers 8010 and 8011. Of course, the effect of datum shift will be visible only after G56 becomes the current WCS.

```
O8010 (G56 DATUM SHIFT ON MILL M/C);
#5261 = #5021 - #5201;
#5262 = #5022 - #5202;
M30;

O8011 (G56 Z-DATUM SHIFT ON LATHE);
#5262 = #5022 - #5202;
M30;
```

The given program may appear trivial (as one can shift the datum manually, using the usual offset setting procedure), but it can be very effectively used. For example, if the workpiece on a lathe is changed to a different one, having different dimensions, the datum setting for Z would need to be repeated (X-datum would remain the same, because the X0 position does not depend on the diameter of the workpiece). If the machine operator is not fully aware of the offset setting procedure, he may simply be asked to first bring the tool manually (in JOG or HANDLE mode) to the desired Z0 position, and then run program number 8009, without disturbing the position of the tool. This would place the Z-datum of the current WCS at the chosen location. Naturally, this has to be done in the same workpiece coordinate system which the machining program uses. If it is, say, G56, one will have to first execute G56 in the MDI mode, before running program number 8009 in AUTO mode. And the RESET key must not be pressed in-between, otherwise, the default workpiece coordinate system, G54, will again become the active coordinate system!

At this stage, it is appropriate to point out a major difference in the way a program is run in MDI and AUTO modes. While these two modes are for different purposes: the MDI mode for 2–3 line programs which are not required to be saved, and the AUTO mode for running long programs stored in the memory of the CNC, this is not the only difference. **When a program is run in the MDI mode, the system does not get reset automatically when the execution is over.** If reset is desired, M02 or M30 will have to be explicitly programmed. On the other hand, **even if M02 or M30 is missing at the end of the program being run in the AUTO mode, the system will get reset automatically when the execution ends** (if parameter 3404#6 is set to 0, the system enters the alarm state if M30 or M02 is missing at the end of the program. The default setting of this parameter is 1, which resets the system without any alarm, even if M30 or M02 is missing). This means that if a single-line program G56 is executed in the MDI mode, G56 will become the active WCS. On the other hand, if the same program is executed in the AUTO mode, the active WCS will change back to G54 at the end of the execution! Another difference, which is not relevant to our discussion, is that *dynamic graphic display* is not available in the MDI mode.

The given program for datum shift, after some changes, can also be used for automatic tool offset setting, using *touch probes*, which will drastically reduce the set-up time. This aspect is discussed in detail in Chap. 11.

5.6   Nested WHILE Statement

The discussion in this chapter is complete, except that no example for a nested WHILE statement has been given, so far. A simple WHILE statement is used when a number of similar operations are required to be done, as long as a specified condition remains satisfied. On the other hand, if each operation involves another set of similar operations, in a conditional loop, then a nested WHILE statement is used. Though the control allows one more level of nesting, it is practically never needed.

The regular rectangular array of holes on a plate, shown in Fig. 5.13, can be very conveniently made using a canned cycle for drilling (say, G81), as all canned cycles can be repeated specified number of times. Program number 8012 is one such program for this job. It is assumed that the required depth of holes is 5 mm, and the top surface of the plate is the datum for the Z-axis. XY-datum is placed at the center of hole-1, as shown in the figure.

Program number 8013 uses a two-level nested WHILE statement, based on the flowchart given in Fig. 5.14. In the nested WHILE statement, the outer loop has been used to drill the required number of holes along the X-axis (i.e., on the Y = 0 line), and the inner loop has been used to create a column of holes at each hole made by the outer loop.

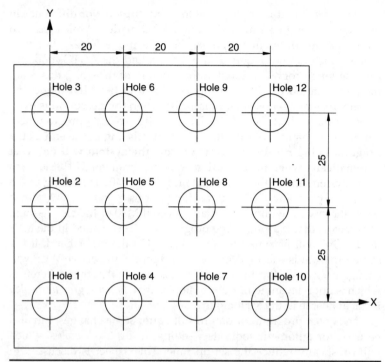

Figure 5.13 A regular rectangular array of holes on a plate.

```
O8012 (HOLE ARRAY USING A CANNED CYCLE);
G21 G94 G90 G54;          (Initial settings)
M06 T1;                   (Tool number 1 placed in the spindle)
M03 S1000;                (Clockwise rpm 1000 starts)
G43 H01;                  (Tool length compensation invoked)
G00 X0 Y0 Z100;           (Tool placed 100 mm above hole 1)
G81 G99 Z-5 R2 F30;       (Hole 1 drilled. Retraction to Z = 2 mm)
G91 Y25 K2;               (Hole 2 and 3 drilled)
X20 Y-50;                 (Hole 4 drilled)
Y25 K2;                   (Hole 5 and 6 drilled)
X20 Y-50;                 (Hole 7 drilled)
Y25 K2;                   (Hole 8 and 9 drilled)
X20 Y-50;                 (Hole 10 drilled)
Y25 K2;                   (Hole 11 and 12 drilled)
G80;                      (Canned cycle cancelled)
M05;                      (Spindle stops)
G90 G00 Z100;             (Tool retracted to Z = 100 mm)
M30;                      (Program end and control reset)


O8013 (HOLE ARRAY USING A NESTED WHILE);
(#1 @ COUNTER1, HOLE COUNTER ALONG X);
(#2 @ COUNTER2, HOLE COUNTER ALONG Y);
```

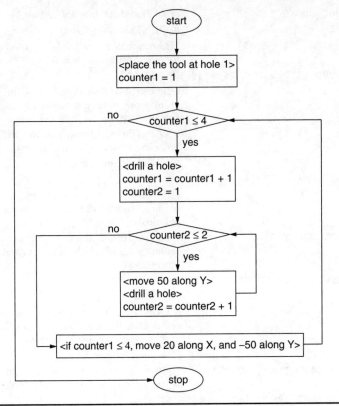

Figure 5.14 Flowchart for an algorithm to drill holes on the plate shown in Fig. 5.13, using a nested WHILE statement.

```
(#3 @ TOTAL NUMBER OF HOLES ALONG X);
(#4 @ TOTAL NUMBER OF HOLES ALONG Y);
(#5 @ DEPTH OF HOLE);
(#6 @ PITCH ALONG X);
(#7 @ PITCH ALONG Y);
;
#3 = 4;                    (Four holes along the X-axis to be made)
#4 = 3;                    (Three holes along the Y-axis to be made)
#5 = 5;                    (Depth of hole is 5 mm)
#6 = 20;                   (Pitch along a row is 20 mm)
#7 = 25;                   (Pitch along a column is 25 mm)
;
G21 G94 G90 G54;           (Initial settings)
M06 T1;                    (Tool number 1 placed in the spindle)
M03 S1000;                 (Clockwise rpm 1000 starts)
G43 H01;                   (Tool length compensation invoked)
```

`G00 X0 Y0 Z100;`	(Tool placed 100 mm above hole 1)
`#1 = 1;`	(COUNTER1 initialized)
`WHILE [#1 LE #3] DO 1;`	(The outer loop starts. It is executed once for every column of holes)
`G90 G81 G99 Z-[ABS[#5]] R2 F30;`	
	(The use of ABS function makes the program work even if a negative value is specified for the depth of the hole)
`#1 = #1 + 1;`	(COUNTER1 incremented for drilling holes in the next column of holes)
`#2 = 1;`	(COUNTER2 initialized)
`WHILE [#2 LE [#4 - 1]] DO 2;`	
	(The inner loop starts. It makes the required number of holes in each column of holes)
`G91 G00 Y#7;`	(The tool shifted along the Y-axis to the center of the next hole in the current column of holes)
`G90 G81 G99 Z-[ABS[#5]] R2 F30;`	
	(Hole drilled)
`#2 = #2 + 1;`	(COUNTER2 incremented for counting and drilling the required number of holes in the current column of holes)
`END 2;`	(The inner loop ends)
`IF [#1 GT #3] GOTO 10;`	(Shift to the base of the next column of holes is not needed after the last column)
`G91 G00 X#6 Y-[#7 * [#4 - 1]];`	
	(Tool shifted to the base of the next column of holes)
`N10;`	(A blank NC statement)
`END 1;`	(The outer loop ends)
`G80;`	(Canned cycle cancelled)
`M05;`	(Spindle stops)
`G90 G00 Z100;`	(Tool retracted to Z = 100 mm)
`M30;`	(Program end and control reset)

While it may appear that the use of the WHILE statement in this case is too complex a method for too simple a problem, this method would definitely be preferable for a large number of holes. In fact, **even if the number of holes increases, the program remains essentially the same**; only the two variables (#3 and #4) for the number of holes along the X- and Y-directions would need to be redefined, as per requirement. While the use of a loop, especially a nested loop, makes a simple program complex, it does make a complex problem pretty simple! Anyway, the basic purpose here was to explain the use of the nested WHILE statement. Moreover, this happens to be our first program which involves machining, and which is also a common practical application.

The usefulness of a flowchart is more than apparent in this example. Even though the flowchart looks simple, and its logic can be very easily verified, this is not so with the program. A program using macro features always looks much more complex than it actually is. So, without a flowchart, not only is it difficult to interpret the program, logical errors also are likely to creep in while developing such a program. On the other hand, if the flowchart is ready and verified, the program coding is just mechanical work.

Note that the given program is more general than the flowchart. This is another trick to make our life simple. First, make a simple flowchart for a special case, and write the program. Once the program is ready, it is easy to modify it for a general case. The given program would accept any values for the number of rows and columns of holes, depth of holes, pitch along rows, and pitch along columns. In fact, it is possible to introduce more complexity in the program. For example, depending on the specified depth of holes, the program can be made to automatically select an appropriate drilling cycle, for example, G81 for shallow holes, G73 for deep holes, and G83 for very deep holes. For this, one simply has to replace G81 by G#8, and define #8 (which can be 81, 73, or 83) on the basis of #5 (depth of hole), using conditional statements in the following manner:

```
IF [#5 LE 5] THEN #8 = 81;      (Shallow holes, depth ≤ 5 mm)
IF [[#5 GT 5] AND [#5 LE 50]] THEN #8 = 73;
                                (Deep holes, 5 mm < depth ≤
                                50 mm)
IF [#5 GT 50] THEN #8 = 83;     (Very deep holes, depth > 50 mm)
G90 G#8 G99 Z-[ABS[#5]] R2 F30;
```

It is also possible to insert error traps in the program, to abort or temporarily halt the execution in case of out-of-range or illegal input data/calculated values:

```
IF [#5 GT 100] THEN #3000 = 1 (HOLE TOO DEEP);
IF [[#6 LE 5] OR [#7 LE 5]] THEN #3006 = 1 (PITCH TOO
SMALL);
```

Note that the first statement would generate an alarm condition, which will terminate further execution of the program. The second statement will only pause the execution, which can be restarted by pressing CYCLE START again, if the operator feels that the specified pitch is correct. In both cases, the messages in the brackets will be displayed on the alarm screen/message screen.

In general, modifying a program is much simpler than preparing the basic program. So, no matter how complex a requirement is, always start with a simplified case. Never worry about the ornamental aspects of a program in the beginning. In fact, once the basic program is ready and works well, the programmer himself gets encouraged to

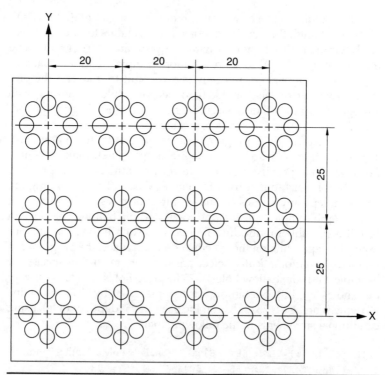

FIGURE 5.15   A part requiring three levels of nested WHILE statement.

make it better! So, as a rule of thumb, always start with a small and easily achievable target.

Coming back to the nesting issue, the reader must have realized by now that had there been a pattern of holes instead of single hole at each location, the third level of nesting would have been required to make the hole-pattern at each location. Such an example is shown in Fig. 5.15. However, three levels of nesting make the logic too complex to understand easily. So, not more than two levels should be used, unless it is absolutely necessary. This would practically always be possible, though the program would not be as compact and general as with three levels of nesting. For example, it is possible to use just one level of nesting (actually, one level of nesting implies that there is no nesting at all) for the job of Fig. 5.15. Just write a macro for the polar pattern (which will require a single level of WHILE statement, without any nesting), and call it 12 times after placing the tool at the centers of all the polar patterns one by one. Such a program will, of course, not be a general-purpose program.

CHAPTER 6

Review of Subprograms

6.1 Introduction

It is necessary to understand why and how subprograms are used before trying to learn how to write macros. Though it is expected that the readers are well aware of subprograms, this chapter presents a brief review to refresh the memory. Even if one has used subprograms before, it is highly recommended that this chapter be read carefully, to eliminate the possibility of any missing link.

What Is a Subprogram?

As the name suggests, a subprogram is a part of a program that is stored in the memory of CNC like any other program with a specified program number. It is not a complete program, and is designed to be called by some other program, which can be a main program, another subprogram, or a macro. A subprogram by itself would generally be meaningless, unless used inside some other program.

Why Is It Used?

Sometimes a program contains repetitive program blocks that appear at several places without any change. For example, assume that a groove of certain geometry and 10-mm depth is to be cut on a flat plate on a milling machine. If the maximum permissible depth of cut is 2 mm, the groove would have to be milled in five passes, after placing the tool at successively increasing depths of 2, 4, 6, 8, and 10 mm. This means that the program blocks, which define the geometry of the groove, would appear at five places in the program, making the program unnecessarily long. In such cases, a better way would be to store the repetitive blocks as a subprogram, and call it five times.

Subprograms versus Subroutines of a Conventional Computer Language

The only similarity between the two is that both are called by other programs. Subprograms are not as flexible and versatile as subroutines. The subroutines of a conventional computer language such as FORTRAN are more like macros that are described in Chap. 7. A subprogram call is simply a *copy-and-paste* operation of the called program into the calling program. Despite its limited scope, use of subprograms does simplify programming, and makes the program tidy.

6.2   Subprogram Call

The syntax of a subprogram call is

```
M98 P1234;
```

which calls program O1234 once, as a subprogram. The program number of a subprogram can be any valid program number, that is, any program in the range O0001 (which is the same as O1, O01, or O001, that is, program number 1) to O9999:

```
M98  P0001;      (Calls program O0001 once)
M98  P001;       (Same as above)
M98  P01;        (Same as above)
M98  P1;         (Same as above)
M98  P9999;      (Calls program O9999 once)
```

The end of subprogram execution is indicated by M99 that returns the execution to the block following the calling block of the program that called the subprogram (i.e., to the block next to M98 P1234, in the first example). M99 is similar to the RETURN statement of FORTRAN.

As an example of using subprograms, consider the triangular groove of Fig. 6.1. Program O0001 would machine this groove by moving a slotdrill of 6-mm diameter along the center line of the groove. The repetitive blocks, which define the geometry of the groove, appear at five places in this program. These blocks have been highlighted for easy identification.

The workpiece zero point has been chosen to lie at the lower left corner of the triangle, as shown in the figure. The Z-datum, for this program as well as for all the programs that appear in this book, is placed at the upper surface of the workpiece.

```
O0001;                   (Program number 1)
G21 G94;                 (Millimeter mode and feedrate in millimeter
                          per minute)
G54;                     (Workpiece coordinate system)
G91 G28 X0 Y0 Z0;        (Reference point return)
```

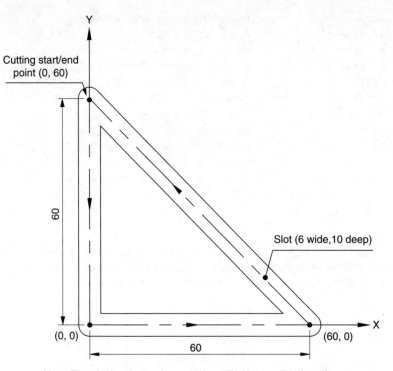

Figure 6.1 Triangular groove on a plate.

Note: The slot has been chosen to be milled in the direction of arrows shown on the center line of the slot. It is also possible to choose some other start/end point, and clockwise direction of machining. The workpiece zero point also can be chosen to lie at some other location.

M06 T1;	(Tool number 1)
M03 S1000;	(CW rpm 1000)
G90 G43 H01;	(Tool length compensation)
G00 X0 Y60 Z100;	(Rapid positioning to 100 mm above the top corner of the groove)
Z1;	(Rapid positioning to 1 mm above the workpiece)
G01 Z-2 F10;	(Hole of 2-mm depth)
Y0 F60;	(2-mm-deep groove along the Y-axis)
X60;	(2-mm-deep groove along the X-axis)
X0 Y60;	(2-mm groove at 45°)
Z-4 F10;	(Hole of 4-mm depth)
Y0 F60;	(4-mm-deep groove along the Y-axis)
X60;	(4 mm-deep groove along the X-axis)
X0 Y60;	(4-mm groove at 45°)
Z-6 F10;	(Hole of 6-mm depth)
Y0 F60;	(6-mm-deep groove along the Y-axis)

X60;	(6-mm-deep groove along the X-axis)
X0 Y60;	(6-mm groove at 45°)
Z-8 F10;	(Hole of 8-mm depth)
Y0 F60;	(8-mm-deep groove along the Y-axis)
X60;	(8-mm-deep groove along the X-axis)
X0 Y60;	(8-mm groove at 45°)
Z-10 F10;	(Hole of 10-mm depth)
Y0 F60;	(10-mm-deep groove along the Y-axis)
X60;	(10-mm-deep groove along the X-axis)
X0 Y60;	(10-mm groove at 45°)
G00 Z100;	(Rapid retraction to 100 mm above the workpiece)
M05;	(Spindle stop)
M30;	(Execution end and control reset)

The program given above can be made shorter by using a subprogram to store the repetitive blocks. Program O0002 is such a program that calls program O0003 as a subprogram, five times.

O0002;	(Program number 2)
G21 G94;	(Millimeter mode and feedrate in millimeter per minute)
G54;	(Workpiece coordinate system)
G91 G28 X0 Y0 Z0;	(Reference point return)
M06 T1;	(Tool number 1)
M03 S1000;	(CW rpm 1000)
G90 G43 H01;	(Tool length compensation)
G00 X0 Y60 Z100;	(Rapid positioning to 100 mm above the top corner of the groove)
Z1;	(Rapid positioning to 1 mm above the workpiece)
G01 Z-2 F10;	(Hole of 2-mm depth)
M98 P0003;	(Subprogram call of O0003)
G01 Z-4 F10;	(Hole of 4-mm depth)
M98 P0003;	(Subprogram call of O0003)
G01 Z-6 F10;	(Hole of 6-mm depth)
M98 P0003;	(Subprogram call of O0003)
G01 Z-8 F10;	(Hole of 8-mm depth)
M98 P0003;	(Subprogram call of O0003)
G01 Z-10 F10;	(Hole of 10-mm depth)
M98 P0003;	(Subprogram call of O0003)
G00 Z100;	(Rapid retraction to 100 mm above the workpiece)
M05;	(Spindle stop)
M30;	(Execution end and control reset)
O0003;	(Program number 3)

Y0 F60;	(Groove along the Y-axis, at the current depth)
X60;	(Groove along the X-axis, at the current depth)
X0 Y60;	(Groove at 45°, at the current depth)
M99;	(Return to the calling program)

6.3   Multiple Call of a Subprogram

It is possible to call the subprogram a desired number of times repeatedly, that is, one after another. Up to 999 repetitions are permitted. The argument of P in M98 block must contain seven digits (or minimum five digits) for this. Four digits from the right are considered the (sub)program number, and the remaining digits (maximum three) to the left indicate the number of repetitions. And if eight digits are specified in the P-word, the eighth digit from the right is **ignored**:

M98 P50001;	(Calls program number 1, five times, which is equivalent to five **consecutive** blocks of M98 P0001 or M98 P1)
M98 P050001;	(Same as above)
M98 P0050001;	(Same as above)
M98 P9990001;	(Program number 1 is called 999 times)
M98 P12340001;	(Program number 1 is called 234 times. The eighth digit from right is ignored)

A subroutine also can be repeatedly called by inserting an L-word in the M98 block. The argument of the L-word is the number of repetitions. The advantage of using this method is that up to 9999 repetitions would be possible, compared to only 999 with the earlier method:

M98 P0001 L9999;	(Program number 1 is called 9999 times)
M98 P1 L9999;	(Same as above)
M98 P1 L10000;	(Illegal L-word)

If the repeated call of a subprogram simply retraces the toolpath, the repetition would be unnecessary and meaningless. Repetition would be useful only in incremental mode where it is possible to have different toolpaths in the subsequent executions of the same subprogram. For example, program O0004 repeatedly calls program O0005 five times in incremental mode, for making the groove of Fig. 6.1.

O0004;	(Program number 4)
G21 G94;	(Millimeter mode and feedrate in millimeter per minute)
G54;	(Workpiece coordinate system)
G91 G28 X0 Y0 Z0;	(Reference point return)
M06 T1;	(Tool number 1)
M03 S1000;	(CW rpm 1000)
G90 G43 H01;	(Tool length compensation)

```
G00 X0 Y60 Z100;    (Rapid positioning to 100 mm above the top cor-
                     ner of the groove)
Z1;                 (Rapid positioning to 1 mm above the workpiece)
G01 Z0 F10;         (The tool touches the workpiece)
G91;                (Incremental coordinate mode)
M98 P50005;         (Subprogram call of O0005, five times)
G90 G00 Z100;       (Rapid retraction to 100 mm above the workpiece)
M05;                (Spindle stop)
M30;                (Execution end and control reset)


O0005;              (Program number 5)
Z-2 F10;            (Depth of hole increased by 2 mm)
Y-60 F60;           (Groove along the Y-axis, at the current depth)
X60;                (Groove along the X-axis, at the current depth)
X-60 Y60;           (Groove at 45°, at the current depth)
M99;                (Return to the G90 G00 Z100 block of the calling
                     program, after five successive executions of the
                     subprogram)
```

Though the subprogram given above works fine, it can be made a little bit more general by invoking the incremental mode inside it. Then, even if it is called in absolute mode, the groove would be correctly made. This reduces the possibility of its incorrect use. Programs O0006 and O0007 use this idea.

```
O0006;              (Program number 6)
G21 G94;            (Millimeter mode and feedrate in millimeter
                     per minute)
G54;                (Workpiece coordinate system)
G91 G28 X0 Y0 Z0;   (Reference point return)
M06 T1;             (Tool number 1)
M03 S1000;          (CW rpm 1000)
G90 G43 H01;        (Tool length compensation)
G00 X0 Y60 Z100;    (Rapid positioning to 100 mm above the top
                     corner of the groove)
Z1;                 (Rapid positioning to 1 mm above the
                     workpiece)
G01 Z0 F10;         (The tool touches the workpiece)
M98 P50007;         (Subprogram call of O0007, five times)
G90 G00 Z100;       (Rapid retraction to 100 mm above the
                     workpiece)
M05;                (Spindle stop)
M30;                (Execution end and control reset)


O0007;              (Program number 7)
G91 Z-2 F10;        (Depth of hole increased by 2 mm)
Y-60 F60;           (Groove along the Y-axis, at current depth)
```

```
X60;                    (Groove along the X-axis, at current depth)
X-60 Y60;               (Groove at 45°, at current depth)
M99;                    (Return to the G90 G00 Z100 block of the call-
                        ing program, after five successive executions
                        of the subprogram)
```

The subprogram given above is definitely better, since it can be called in both absolute and incremental modes. Nevertheless, it does have a shortcoming: it forces incremental mode when the execution returns to the calling program. Moreover, it assumes that G01 is active at the time of calling it. This is yet another source of programming errors. Subprograms and macros should be made as general as possible/practical, because usually these are developed by expert programmers for use in several programs prepared by other program-mers. If a new programmer calls a subprogram developed by some other programmer, he definitely does not expect that the subprogram would change the coordinate mode or any other setting of the calling program! Fortunately, using macro-programming features, it is pos-sible to identify the current coordinate mode, among several other things. So, a subprogram can be designed in such a manner that it does not change the current status of the control. This can be done by identifying the status of the control in the beginning of the subpro-gram, and restoring it in the end, that is, just before exiting to the calling program by M99.

Note that the subprogram O0007 needs to use three things: G01, G91, and F-code. G01 belongs to G-code Group 01, and G91 belongs to Group 03. The other G-codes in Group 01 are G00, G02, G03, and G33 on a milling machine. On the other hand, G90 is the only other G-code in Group 3. Referring to Table 3.10(b), the system variable #4001 stores 0, 1, 2, 3, or 33 depending on the currently active G-code of Group 01 (G00, G01, G02, G03, and G33, respectively). For exam-ple, if G01 is currently active, #4001 would store 1 (recall that, at any time, one G-code from each group remains active). Similarly, system variable #4003 stores 90 or 91 depending on which one of G90 and G91 is currently active. The current feedrate is stored in system vari-able #4109.

The subprogram O0008 uses these system variables to ascertain the current status of the control in the beginning of the program, and restores it in the end (recall that a statement such as G01, without any axis words, simply changes the currently active G-code of Group 1 to G01, without causing any tool movement). Note that the logic works even in the case of multiple calls of the subprogram.

In addition, for the purpose of making subprogram O0008 even more general, spindle speed also can be specified inside it, so that a possibly incorrect rpm specified in the calling program is not used. Since the subprogram assumes a constant rpm, G97 would need to be specified. Referring to Table 3.10(b), system variable #4013 stores 96 or

97 depending on which one of G96 (constant surface speed) and G97 (constant rpm) is currently active, and the current spindle speed/constant surface speed is stored in the system variable #4119.

Moreover, the feedrate in the subprogram is desired to be in millimeter per minute. So, G94 is also needed. Referring to Table 3.10(b) again, system variable #4005 stores 94 or 95, corresponding to G94 and G95.

Finally, the subprogram has been written in millimeter mode. So, if it is called in inch mode, the execution should be terminated with an alarm message (recall that the coordinate mode cannot be switched in the middle of a program). System variable #4006 stores 20 or 21, corresponding to G20 and G21 [refer to Table 3.10(b).]

O0008;	(Program number 8)
#1 = #4001;	(#1 stores 0, 1, 2, 3, or 33)
#2 = #4003;	(#2 stores 90 or 91)
#3 = #4109;	(#3 stores the current value specified in the F-word, i.e., the current feedrate)
#4 = #4013;	(#4 stores 96 or 97)
#5 = #4119;	(#5 stores the current value specified in the S-word, i.e., the current rpm or the constant surface speed, depending on which one of G97 and G96 is currently active)
#6 = #4005;	(#6 stores 94 or 95)
#7 = #4006;	(#7 stores 20 or 21)
IF [#7 EQ 20] THEN #3000 = 1 (INCH MODE NOT PERMITTED);	(If G20 is used in the calling program, the execution would immediately terminate with the alarm message "3001 INCH MODE NOT PERMITTED." Refer to Macro Alarms in Sec. 3.5)
G94 G97 G91;	(Selects feedrate in millimeter per minute, constant rpm and incremental mode)
G01 Z-2 F10 S1000;	(Depth of hole increased by 2 mm at 1000 rpm and 10 mm/min feedrate)
Y-60 F60;	(Groove along the Y-axis, at the current depth)
X60;	(Groove along the X-axis, at the current depth)
X-60 Y60;	(Groove at 45°, at the current depth)
G#1 G#2 F#3 G#4 S#5 G#6;	(Restores the original control status)
M99;	(Return to the calling program)

The subprogram O0008 is fairly general. It only assumes that the tool (a 6-mm-diameter slotdrill) touches the workpiece at the top corner of the triangular groove, at the time of calling it. The only limitation is that it can only make grooves of depths in the multiple of 2 mm. For example, M98 P40008 would make a groove of 8-mm depth. But it is not possible to have a depth of, say, 5 mm, as it is not a multiple of 2 mm. Moreover, the tool remains at the bottom of the top corner of the groove when the execution returns to the calling program. Though the XY-position is the same as in the beginning, the Z-position also should be restored. Ideally, a subprogram or a macro should do whatever it is required to do, without disturbing any control setting.

Though macros can be made to do virtually anything, subprograms have their own limitations, because it is not possible to pass on any values to its variables while calling it. Moreover, a subprogram does not have its own set of local variables. It only acts as an **extension** of the calling program.

A macro for an arbitrary depth of the groove is given in Chap. 8. If the same thing is to be done through a subprogram, a variable, say, #100 will need to be defined, storing the desired depth of the groove, before calling the subprogram, so that the subprogram could use it. Since such a subprogram uses a logic involving execution in a loop, a flowchart of the chosen algorithm must be prepared before writing the program. Figure 6.2 shows such a flowchart. Subprogram O0009, which is based on this flowchart, is given along with the relevant part of the calling program. Note that O0009 needs to be called only once, for any given depth of groove. Use of loops eliminates the need for multiple calls of subprograms.

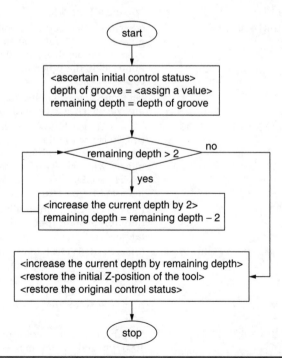

Figure 6.2 Flowchart of an algorithm for making a groove of any specified depth, using a WHILE statement.

```
(Calling program)
...
...
#100 = 5;                    (Specify the desired depth of the
                              groove)
M98 P0009;                   (Subprogram call of O0009)
...
...
M30;                         (Execution end and control reset)

O0009;                       (Program number 9)
#1 = #4001;                  (#1 stores 0, 1, 2, 3, or 33)
#2 = #4003;                  (#2 stores 90 or 91)
#3 = #4109;                  (#3 stores the current value specified in
                              the F-word, i.e., the current feedrate)
#4 = #4013;                  (#4 stores 96 or 97)
#5 = #4119;                  (#5 stores the current value specified in
                              the S-word, i.e., the current rpm or the
                              constant surface speed, depending on
                              which one of G97 and G96 is currently
                              active)
#6 = #4005;                  (#6 stores 94 or 95)
#7 = #4006;                  (#7 stores 20 or 21)
#8 = ABS[#100];              (#8 stores the desired depth of the
                              groove. The ABS function ensures that
                              even if the depth is specified with a
                              minus sign, the program would work
                              correctly)
#9 = #8;                     (#9 stores the remaining depth of the
                              groove, which is to be machined)
IF [#7 EQ 20] THEN #3000 = 1 (INCH MODE NOT PERMITTED);
                              (If G20 is used in the calling program,
                              the execution would immediately ter-
                              minate with the alarm message "3001
                              INCH MODE NOT PERMITTED."
                              Refer to Macro Alarms in Sec. 3.5)
G94 G97 G91;                 (Selects feedrate in millimeter per
                              minute, constant rpm, and incremen-
                              tal mode)
WHILE [#9 GT 2] DO 1;        (Start of the WHILE loop. The loop is
                              executed if the remaining depth of
                              the groove is more than 2 mm. If not,
                              the loop is skipped, and the execu-
                              tion jumps to the block following
                              END 1)
G01 Z-2 F10 S1000;           (Depth of hole increased by 2 mm at
                              1000 rpm and 10 mm/min feedrate)
Y-60 F60;                    (Groove along the Y-axis, at the current
                              depth)
```

`X60;`	(Groove along the X-axis, at the current depth)
`X-60 Y60;`	(Groove at 45°, at the current depth)
`#9 = #9 - 2;`	(Remaining depth recalculated)
`END 1;`	(End of the WHILE loop. The execution jumps to the start of the loop to check if the remaining depth of the groove is still more than 2 mm)
`G01 Z-#9 F10;`	(Depth of hole increased by the required amount, which is now less than or equal to 2 mm, to reach the final depth of the groove, at 1000 rpm and 10 mm/min feedrate)
`Y-60 F60;`	(Groove along the Y-axis, at the final depth)
`X60;`	(Groove along the X-axis, at the final depth)
`X-60 Y60;`	(Groove at 45°, at the final depth)
`G00 Z#8;`	(Restores the initial Z-position of the tool)
`G#1 G#2 F#3 G#4 S#5 G#6;`	(Restores the original control status)
`M99;`	(Return to the calling program)

In program O0009, it is assumed that the maximum permissible depth of cut is 2 mm. As an exercise, try to modify the program when the maximum permissible depth of cut is 1 mm. Also, try to construct the loop using the IF_GOTO_ statement. The flowchart for such a case is given in Fig. 6.3. Note, however, that the WHILE statement is preferred over the IF_GOTO_ statement for constructing loops, as discussed in Chap. 5.

The flowchart given in Fig. 6.3 would need to use the IF_GOTO_ statement twice. This is because the desired depth of the groove may even be less than 1 mm. In the loop created by the IF_GOTO_ statement, the loop terminating condition is checked in the end. Therefore, it is also necessary to check the desired depth before entering the loop, to decide whether the loop should be executed. This problem was not there with the WHILE loop because the loop terminating condition is checked in the beginning of the loop. This is yet another reason why the WHILE loop is preferred over the IF_GOTO_ loop.

Programs O0008 and O0009 have used several of the macro-programming features. The use of macro-programming features is not restricted to developing macros only, even though this is their main purpose. The main programs also can use these features. In fact, this is the way one should start learning macro programming. Writing macros for different applications should be started only after mastering the basic features of the language, by using these in main programs and subprograms, as much as possible.

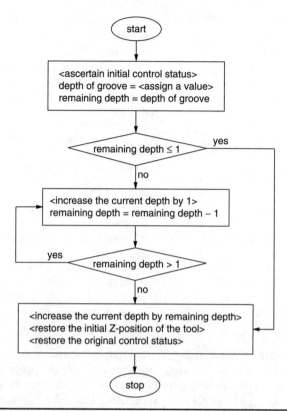

Figure 6.3 Flowchart of an algorithm for making a groove of any specified depth, using GOTO statements.

6.4 Subprogram Nesting

A subprogram can call another subprogram, which is called nesting. A maximum of four levels of nesting, involving the main program and four different subprograms, is permitted, that is, the main program can call subprogram 1, subprogram 1 can call subprogram 2, subprogram 2 can call subprogram 3, and subprogram 3 can call subprogram 4 (subprogram 4 cannot call any other subprogram). This is pictorially represented in Fig. 6.4, where the flow of execution is shown by arrows. Note that there is no restriction on the total number of subprograms or the number of nested subprograms in a main program, only that the level of nesting **at a given place** should be four or less.

Refer also to Fig. 3.2 that explains that the variables of the main program and all the nested subprograms belong to the same set of variables. For example, if a variable, say #1, is being used by the main program as well as all the nested subprograms, then all five occurrences of #1 refer to the same memory location.

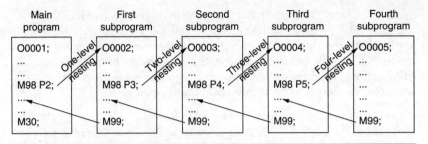

Main program	First subprogram	Second subprogram	Third subprogram	Fourth subprogram
O0001;	O0002;	O0003;	O0004;	O0005;

FIGURE **6.4** Subprogram nesting.

As an example of subprogram nesting, consider the job shown in Fig. 6.5, where five triangular grooves of 10-mm depth are to be machined on a plate of 600 mm × 300 mm size. The nesting feature of subprograms can be used to make the program for such a **regular and repetitive pattern** very short. Once again, the maximum permissible depth of cut is assumed to be 2 mm. Therefore, since the depth of the grooves is 10 mm, five passes for each triangle would be required. The whole exercise would need to be repeated five times, for making five triangular grooves. This can be done by two-level nesting. The call of subprogram O0011 makes the triangular grooves at five places, and the nested call of subprogram O0012 makes each groove in five

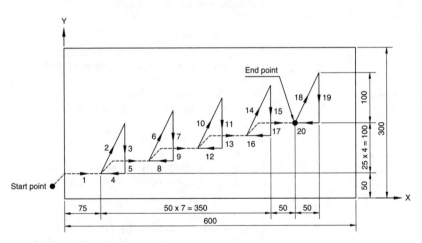

Note:
1. The center lines of the five triangular grooves are shown. The width of the grooves, which is not shown, would be equal to the tool diameter.
2. All the triangles are of the same size, and are evenly spaced.
3. The sequence of toolpath segments is numbered, and the cutting direction is shown by arrows. Due to multiple passes, the toolpath is 1→ (2 → 3 → 4) five times → 5 → (6 → 7 → 8) five times. → 9..., till the end point.
4. The dashed lines represent rapid motion with G00, 1 mm above the workpiece, showing *dog-leg effect*. Rapid positioning in a straight line is also possible, if parameter 1401#1 is set to 1.

FIGURE **6.5** Five triangular grooves in a regular pattern.

passes. The workpiece zero point is chosen to lie at the lower left corner of the plate, with Z-datum at its upper surface. Note that, in such applications, the subprograms need to be written in the incremental mode. No attempt has been made in these programs to make these suitable for a general case, using macro-programming features, as the sole objective here is to illustrate subprogram nesting.

O0010;	(Program number 10)
G21 G94;	(Millimeter mode and feedrate in millimeter per minute)
G54;	(Workpiece coordinate system)
G91 G28 X0 Y0 Z0;	(Reference point return)
M06 T1;	(Tool number 1)
M03 S1000;	(CW rpm 1000)
G90 G43 H01;	(Tool length compensation)
G00 X-25 Y25 Z100;	(Rapid positioning to 100 mm above the start point, as designated in Fig. 6.5)
Z1;	(Rapid to 1 mm above the workpiece)
M98 P50011;	(Subprogram call of O0011, five times, for the five triangles)
G00 Z100;	(Rapid retraction to 100 mm above the workpiece)
M05;	(Spindle stop)
M30;	(Execution end and control reset)
O0011;	(Program number 11)
G91 G00 X100 Y25;	(Rapid positioning to the lower left corner of the triangular groove that is to be machined next)
G01 Z-1 F10;	(The tool touches the workpiece)
M98 P50012;	(Subprogram call of O0012, five times, for the five passes)
G90 G00 Z1;	(Rapid retraction to 1 mm above the workpiece)
M99;	(Return to G00 Z100 block of O0010, after five successive executions of this subprogram)
O0012;	(Program number 12)
G91 G01 Z-2 F10;	(Depth of hole increased by 2 mm)
X50 Y100 F60;	(Groove cutting along the slant side of the triangle, at the current depth)
Y-100;	(Groove cutting along the Y-axis side of the triangle, at the current depth)
X-50;	(Groove cutting along the X-axis side of the triangle, at the current depth)
M99;	(Return to G90 G00 Z1 block of O0011, after five successive executions of this subprogram)

The example given above uses two-level nesting. Though the control permits nesting up to four levels, nesting beyond two levels is rarely used, as such programs tend to become too complex to understand. The objective of nesting is to simplify programming, not to complicate it!

Though nesting and use of macro-programming features in subprograms make a program quite general and compact, the real power comes through macros that have virtually unlimited scope. Of course, a thorough understanding of subprograms is absolutely necessary for developing professional-quality macros. Therefore, the readers should read and re-read each and every line of this chapter, until the concepts are crystal clear.

CHAPTER 7

Macro Call

7.1 Introduction

Several of the macro-programming features have already been discussed. The use of these features is not restricted to just macros. They can be used anywhere—in the main program, subprograms, and macros. A macro program, however, is a unique feature of macro programming, which adds an entirely new dimension to what can be done through programming. Developing a macro does require good programming skills, possessed by very few individuals, but using a macro is pretty simple—it is like commanding a canned cycle. This chapter explains how to define and use macros.

7.2 Macro versus Subprogram

A macro can be broadly described as a sophisticated version of subprograms (this is the reason why this chapter is preceded by the chapter, Review of Subprograms). The overall program structure is the same for both. A subprogram can be called by the main program, other subprograms or macros, and it can also call any type of program. Similarly, a macro can be called by any program, and it can call any program. Both allow nesting up to a maximum of four levels. In case of a mixed nesting, a maximum of four subprograms and four macros can appear, in any order, apart from the main program.

The single major difference between a subprogram and a macro lies in the **flexibility of the input data**. A subprogram either does not use variables at all, or always uses **fixed initial data** for the variables used inside it, with values as defined in the calling program at the time of calling the subprogram. A macro, on the other hand, uses the values specified in the **macro-call statement**, for its local variables. (If the macro call statement does not define a local variable, it remains null, initially.) In fact, as discussed in Chap. 3, a subprogram does not even have its **own** set of local variables. It uses the variables of the calling program. Common variables, permanent common variables, and system variables, of course, behave in the same manner in both subprograms and macros.

7.3   Macro Call

By now, it should be clear that the main difference between a macro call and a subprogram call is that the level of local variables changes with a macro call, but it does not change with a subprogram call. Moreover, a macro call can pass data to the called program, whereas a subprogram call does not have this capability.

A subprogram is called by M98 and M198. A macro can be called using any of the following methods:

- Simple call (G65)
- Modal call (G66)
- Call with user-defined G-code
- Call with user-defined M-code

Those who are new to macro programming may skip the last two methods of calling a macro, at this stage. Even the modal call (G66) is not very frequently used. G65 can be used for calling any macro, without any limitation. Other methods do provide some convenience, but one can manage with G65 also. However, one must know all the methods because of the specific advantages these offer. Moreover, any method may have been used in the programs developed by other programmers, which one can understand only if one knows the method used. However, in the interest of a steeper learning curve, one may choose to read about G65 only. The remaining methods can be learned at a later stage. One may skip Sec. 7.4 also.

Simple Call (G65)

The syntax of macro call with G65 is

```
G65 P<program number> L<repetition count> <argument 1>
... <argument n>;
```

The program number is the number of an existing program that is to be called. Usually, 8000 or 9000 series numbers are used as macro-program numbers, since these can be protected from accidental editing/deletion, through parameters 3202#0 and 3202#4, respectively. However, any legal program number (1 to 9999) can be used.

Repetition count, which defaults to one (i.e., L1 can be omitted), is the number of successive executions of the called macro. This is the only way a macro can be repeatedly executed. The repetition count of a macro cannot be included in the P-word, unlike what can be done for repeating a subprogram call. As explained in Chap. 6, the repetition count of a subprogram can be included in the P-word also (apart from the L-word method). Here, the P-word is used only for specifying the program number. Up to 9999 macro repetitions are possible.

The argument specification is for passing on the desired initial values to the local variables of the macro. All the local variables, #1 through #33, can be initialized. The uninitialized local variables remain null in the beginning of the macro execution. The exact method of argument specification is discussed in Sec. 7.5.

As an example, G65 P8000 L2 A10 B20 calls program number 8000 twice, in succession, with #1 (because of address A) and #2 (because of address B) set to 10 and 20, respectively, whereas the other local variables (#3 through #33) remain null, initially. Note that if the called macro modifies variables #1 and/or #2, its second call would start with the **new values** stored in #1 and #2. Similarly, if the first call defines a new local variable, say, #3, then the second call would start with the stored value in #3. (The first call started with a null value for #3.) **In the repeated calls of a macro, the specified values in the argument list are passed on to the local variables only in the first execution of the macro.** Any subsequent execution uses the values obtained in the previous execution.

Thus, G65 P8000 L2 A10 B20 is **not** equivalent to G65 P8000 A10 B20 commanded twice in two successive blocks. **If a macro is called in two (or more) successive blocks, all the executions become independent of the previous executions, with each execution using the specified values in its macro-call block, as the initial values for its local variables.**

For a better clarity, consider program O0013 that repeatedly calls subprogram O0014 and macro O0015, thrice each, followed by three **separate** macro calls of O0016 in three successive blocks. Program numbers 14, 15, and 16 are essentially same, but behave differently depending on how these are called, which explains the underlying principles. Recall that the same program becomes a subprogram when called by M98, and a macro when called by G65 or G66. Some of these concepts have already been discussed in Chap. 3. (Refer back to Figs. 3.1, 3.2, and 3.3.)

O0013;	(Program number 13)
#1 = 1;	(The value assigned here is used only by this program and the nested subprogram O0014, but not by the nested macros, because a subprogram uses the local variables of the calling program, whereas macros use a different set or level of local variables)
M98 P30014;	(Same as M98 P0014 L3, to execute O0014 thrice in succession, as a subprogram)
G65 P0015 L3 A1;	(Calls O0015 as a macro, thrice in succession. The first execution uses #1 = 1, initially, but the second and the third executions start with the updated values of #1, obtained at the end of the first and the second executions, respectively)
G65 P0016 A1;	(Calls O0016 as a macro once, with #1 set to 1, initially)

`G65 P0016 A1;`	(This macro call is independent of the call in the previous block. So, O0016 is executed once, with #1 again initially set to 1, even if #1 of the previous execution contained a different value. In fact, #1 of the previous execution carries no meaning now)
`G65 P0016 A1;`	(Same as above)
`#503 = #1;`	(Since the subprogram call modified the value stored in #1 to 7, #503 would store 7. A macro call has no effect on the local variables of the calling program)
`M30;`	(Execution end and control reset)
`O0014;`	(Program number 14)
`#1 = #1 + 2;`	(Since this program has been called as a subprogram, #1 used here is the **same** as #1 of the calling program O0013. Since every execution of this program adds 2 to the value stored in #1, the final value becomes 7, after three executions)
`#500 = #1;`	(#500 stores 3 after the first execution, 5 after the second, and 7 after the third and final execution)
`M99;`	(Return to the calling program, after all the specified number of repetitions are over. After the return, #1 of the calling program would contain 7, since it is the same as #1 of this subprogram)
`O0015;`	(Program number 15)
`#1 = #1 + 2;`	(Since this program has been called as a macro, #1 used here is **different** from #1 of the calling program O0013. The macro call assigns a value of 1 to the macro variable #1, in the beginning of the **first** call of this program. The repeated calls use the **updated** values of #1. Since every execution of this program adds 2 to the value stored in #1, the final value becomes 7, after three executions)
`#501 = #1;`	(#501 stores 3 after the first execution, 5 after the second, and 7 after the third and final execution)
`M99;`	(Return to the calling program, after all the specified number of repetitions are over. Since #1 of this program and #1 of the calling program are different variables, #1 of the calling program would still contain 7. It is just a matter of chance that #1 of this program also contains 7 after three executions. In fact, when the execution returns to the calling program, after executing a macro, all the local variables of the macro become **non-existent, hence meaningless**)
`O0016;`	(Program number 16)
`#1 = #1 + 2;`	(Since this program has been called as a macro, **without repetition**, its execution **always** starts with #1 = 1, irrespective of the number of executions of this program)

```
#502 = #1;
```
(#502 stores 3, irrespective of how many times this macro is called)

```
M99;
```
(Return to the calling program)

To summarize, the first execution of the subprogram starts with #1 = 1, and sets #1 = 3 and #500 = 3 when the execution ends. Every subsequent execution of the subprogram adds 2 to these values. Therefore, at the end of three executions of the subprogram, both #1 and #500 store 7. Next comes the macro call of O0015, with the macro variable #1 initially set to 1 (because of the A1 word in the macro call). At the end of macro execution, both #1 and #501 store 3. Now, the macro is to be executed two more times. Since the subsequent executions start with the updated value stored in #1, #501 stores 7 at the end of three calls of the macro. Finally, the three calls of macro O0016 are independent of one another. So, #1, and hence #502 also, store 3 at the end of each execution. In the end, #503 stores the value stored in #1 of the main program. Thus, the result (stored values in the permanent common variables) after complete execution of O0013 is

#500 stores 7.

#501 stores 7.

#502 stores 3.

#503 stores 7.

This example clearly explains that the local variables of the main program and the subprogram are the same (belonging to level 0), but the macro uses a different set of local variables (belonging to level 1). The permanent common variables (#500 and #501) have been used here, so that the values stored in them could be inspected at the end of the program execution. Recall that the content of a permanent common variable is not washed out by system reset or power down. However, this also means that if a program does not modify the content of a permanent common variable, the variable would display the old value stored in it! Therefore, run program number 8001 or 8005, given in Chap. 5, for clearing all permanent common variables, if their old values are likely to cause confusion.

Modal Call (G66)

Sometimes a macro must be called several times, but not at the same tool position. For example, assume that a macro has been written for making a hole in some special manner (e.g., with progressively reducing peck lengths). Now, if several holes are to be drilled, the macro would need to be called, using G65, several times, after bringing the tool to the desired locations:

```
...
...
<Move the tool to location 1>
G65 P_ L_ <argument specification>;
<Move the tool to location 2>
G65 P_ L_ <argument specification>;
<Move the tool to location 3>
G65 P_ L_ <argument specification>;
...
...
<Move the tool to location n>
G65 P_ L_ <argument specification>;
...
...
```

Though there is nothing wrong with this method, the G65 block needs to be inserted repeatedly. In case any change in the G65 block is desired, the change would have to be incorporated in all the G65 blocks. This not only involves extra typing effort, the program size also increases. The modal call, G66, removes this difficulty. A G66 macro call remains effective until canceled by G67. After a G66 block, whenever **the tool is moved** in a subsequent block, the macro is automatically called, **after completing the motion**, till G67 is programmed, which cancels the G66 mode. The program structure with G66 would take the following form (this is similar but not exactly equivalent to the G65-form given above, due to reasons explained below):

```
...
...
G66 P_ L_ <argument specification>;
<Move the tool to location 1>
<Move the tool to location 2>
<Move the tool to location 3>
...
...
<Move the tool to location n>
G67;
...
...
```

The syntax of G66 is exactly same as that of G65. All the rules applicable to G65 apply also to G66. A major difference is that G66 **only stores the information** provided in its block as *modal data* (i.e., data for future use, until changed), **without calling the macro**. The macro is called **only when** the tool is moved in a subsequent block, after the completion of the specified motion. Moreover, the values specified in the G66 block are passed on to the macro **only in its first call**. All other calls, till G67 is encountered or another G66 is commanded, use the **updated** values for the local variables, obtained from the **preceding** execution. Program numbers 17, 18, and 19 illustrate these rules.

Single call of a macro with G66:

O0017;	(Program number 17)
G66 P0015 A1;	(Stores modal information, without calling the macro. Though program number 15 is the same program given earlier, the explanations given on the right-hand side would not apply here)
G91 G00 X100;	(The tool moves 100 mm along the X-axis. Thereafter, O0015 is called as a macro, with #1 initially set to 1. The execution of the macro assigns a value of 3 to both #1 and #501)
X100;	(The tool further moves 100 mm along the X-axis. Thereafter, O0015 is again called as a macro, with #1 initially set to 3, its **updated** value in the previous execution. Therefore, the current execution of the macro assigns a value of 5 to both #1 and #501)
G67;	(G66 mode canceled)
X100;	(Since G66 is no longer active, only the tool would move by 100 mm, without calling any macro)
M30;	(Execution end and control reset)

Repeated call of a macro with G66:

O0018;	(Program number 18)
G66 P0015 L3 A1;	(Stores modal information, without calling the macro)
G91 G00 X100;	(The tool moves 100 mm along the X-axis. Thereafter, O0015 is called as a macro thrice, with #1 **initially** set to 1. The three executions of the macro assign a value of 7 to both #1 and #501)
X100;	(The tool further moves 100 mm along the X-axis. Thereafter, O0015 is again called thrice as a macro, with #1 initially set to 7, its **updated** value in the previous execution. Therefore, the current three executions of the macro assign a value of 13 to both #1 and #501)
G67;	(G66 mode canceled. G67 can be omitted here because the next command is M30, causing control reset, which automatically changes the control status to G67)
M30;	(Execution end and control reset)

Macro call with multiple G66 blocks:

O0019;	(Program number 19)
G66 P0015 A1;	(Stores modal information, without calling the macro)

G91 G00 X100;	(The tool moves 100 mm along the X-axis. Thereafter, O0015 is called as a macro, with #1 initially set to 1. The execution of the macro assigns a value of 3 to both #1 and #501)
G66 P0015 A1;	(Stores modal information, without calling the macro. A second or any subsequent G66 block remains independent of previous G66 blocks. In fact, the previous G66 automatically gets canceled, and the new G66 takes effect. So, the updated values for the local variables from the previous execution are not used. The next execution of O0015 would again start with the specified data in the current G66 block)
X100;	(The tool further moves 100 mm along the X-axis. Thereafter, O0015 is again called as a macro, with #1 initially set to 1, its specified value in the **currently active** G66 block. Therefore, the current execution of the macro assigns a value of 3 to both #1 and #501, which happens to be the same as that obtained in the previous execution of the macro, because all the modal parameters are same in both the G66 blocks)
G67;	(The current G66 mode canceled. The previous G66 mode was automatically canceled when a new G66 was commanded)
M30;	(Execution end and control reset)

Additional comments:

- Nesting with G66 is permitted. A maximum nesting depth of four levels is allowed, including **both** G65 and G66. Subprograms also can be nested to G66. Over all, a maximum of four macros (called by G65 and/or G66) and four subprograms can be nested.

- The G66 block only stores modal data. It **does not** call a macro. Therefore, no movement command should be included in the G66 block. If the movement command (such as X100) appears before the G66 word, the tool moves without calling the macro (though modal data are stored for future use). If it appears after the G66 word, it is taken as data for a local variable (X100 would assign a value of 100 to variable #24)! Moreover, an address such as G01 at the right of G66 is illegal, since the letters L, O, N, G, and P are not allowed to be used for argument specification.

- While G66 is active, M-codes or any other code that does not cause tool movement are executed **without** calling the macro.

- As with G65, up to 9999 repetitions (specified with address L) in a G66 macro call are permitted.

- The modal data are defined **only** in the G66 block. For example, in a subsequent block where G66 is active, X100 L3 would not change the repetition count to 3.

- If a macro call is needed without tool movement, specify zero distance for some axis in incremental mode. For example, G91 G00 X0 (G00 U0 on a lathe) would call the macro without any tool movement. Actually, an **axis movement block** is necessary; the specified displacement can be **zero also**!

- In the *word address format*, which today's CNC machines use, the order of different words associated with a command is not important. For example, G66 P15 A1 and G66 A1 P15 are both equivalent (though P15 G66 A1 is illegal, since all arguments of a G-code must come **after** the code). However, only the recommended order should be used, so as to eliminate any possibility of confusion to other users.

Call with User-Defined G-Code

It is possible to define up to 10 **new** G-codes that can be used for calling a macro in the way G65 calls it, with the difference that the program number is not required to be mentioned with the defined G-code. Such G-codes become similar to the built-in standard G-codes, especially like the codes for canned cycles (those canned cycles that are defined in one block). For using these codes, one only needs to know the meanings of the used letter addresses. For example, G100 D80 Q5 R90 F10 S1000 may be designed to call a macro for drilling a hole with progressively reducing peck lengths, with the following meanings for its letter addresses. (We have not yet discussed the correspondence between various letter addresses and the local variables of the called macro. The only thing we have used so far is letter A defines variable #1 and letter B defines variable #2. So, do not worry about how the called macro in this example uses the given data. It is discussed later.)

`G100:`	New G-code
`D80:`	Depth of hole (80 mm)
`Q5:`	First peck length (5 mm)
`R90:`	Reduction percentage for the next peck length (90%)
`F10:`	Feedrate (10 mm/min)
`S1000:`	Spindle speed (1000 rpm)

If G100 is designed to call macro O9010, the given command is equivalent to G65 P9010 D80 Q5 R90 F10 S1000.

The advantage of calling a macro in this manner is that even a programmer with less programming skills can make use of some (not all) predefined macros, **without** needing to know anything about macro programming! A new G-code can be used just like a standard G-code. One **need not** even know the macro program number it calls. In other words, some **extra** G-codes become available on the machine. Of course, only an experienced programmer can define such codes. The exact procedure is described below.

Program Number	Parameter Number
09010	6050
09011	6051
09012	6052
09013	6053
09014	6054
09015	6055
09016	6056
09017	6057
09018	6058
09019	6059

TABLE 7.1 Correspondence between Program Numbers and Parameter Numbers, for Macro Call Using a G-Code

Only the macros bearing program numbers 9010 through 9019 can be called by this method. The G-codes for calling theses macros can have any number between 1 and 9999. However, only the unused G-codes should be selected, otherwise the existing G-code would get **redefined**!

The G-code numbers, which are selected for calling macros (in the range 9010 to 9019), are specified in certain parameters (6050 through 6059). The correspondence between the two is given in Table 7.1. For example, if 100 is stored in parameter 6050, G100 would call the macro O9010. Similarly, if 1000 is stored in parameter 6059, G1000 would call the macro O9019. And, if the program number corresponding to a defined G-code does not exist, the control would enter the alarm state, displaying the "NUMBER NOT FOUND" alarm message. For example, storing 100 in parameter 6050, without defining program O9010, would cause an alarm state if G100 is commanded.

All the rules and restrictions that apply to G65 also apply to macro call with G-codes. For example, 1 to 9999 repetitions can be specified with the address L. The arguments also are specified in the same manner.

For the purpose of illustration, consider a very simple example of defining a code G500 (without any arguments) to unconditionally cause a dwell of 10 seconds. For this, store 500 in parameter 6050 (assuming that parameter 6050 has not been not used for defining some other G-code). Then, G500 would call macro O9010 that should be defined in the following manner:

O9010;   (The program number corresponding to parameter number
            6050, as per Table 7.1)

G04 X10;   (Causes a dwell of 10 seconds)

M99;   (Return to the calling program)

Now, in any program, wherever G100 is commanded, it would result in a dwell of 10 seconds. And, for a dwell of, say, 50 seconds, G500 L5 would need to be programmed. G65 P9010 and G65 P9010 L5 also would do the same thing, causing delays of 10 and 50 seconds, respectively. Basically, G65 P_ is being replaced by a G-code, for convenience. Extremely complex macros also can be used for defining new G-codes. Such an example is not given here, as the idea was to explain only the methodology.

Apart from this method of calling a program, a program can also be called with M-codes and T-codes, as discussed in the next section. In a program, which is called by any of these methods, all G-codes are treated as ordinary G-codes, with their original meanings. In other words, a G-code in such programs would not call a program, even if it is designed to call a program. It would be executed in the usual manner. This issue is discussed in more detail later.

Call with User-Defined M-Code

The way a macro can be called by a G-code, it can also be called by an M-code, exactly in the same manner. This method of macro call defines a new M-code, or redefines an existing M-code, if the number of a predefined M-code is chosen. Therefore, only an unused number should be selected.

Any number between 1 and 99,999,999 can be used, which would need to be stored in any of the parameters between 6080 and 6089. Up to 10 M-codes can be defined. These call program numbers between O9020 and O9029. The correspondence between the parameter numbers and the program numbers is given in Table 7.2. As an example, if M1000 is desired to be defined, 1000 would need to be stored in, say, parameter 6080 that would make M1000 call the macro O9020. And, if O9020 does not exist, an alarm would be generated, with the message "NUMBER NOT FOUND." It is possible to specify up to 9999 repetitions of the called macro, using the L-word. As in case of a G-code macro call, the method of specifying the arguments in the M1000 (or in any other user-defined M-code) block, for passing on the desired values to the local variables of the called macro, is the same as that used in a G65/G66 macro call.

There is a restriction that the M-code macro call, along with all the arguments, be used alone in its block. Otherwise, it would either be an illegal command, or it may ignore the other codes. Moreover, as in case of macro call using G-codes, this method of macro call cannot be used in programs called by G-codes, M-codes, or T-codes. In such programs, all M-codes possess their usual meanings; no macros are called.

Program Number	Parameter Number
O9020	6080
O9021	6081
O9022	6082
O9023	6083
O9024	6084
O9025	6085
O9026	6086
O9027	6087
O9028	6088
O9029	6089

TABLE 7.2   Correspondence between Program Numbers and Parameter Numbers, for Macro Call Using an M-Code

7.4   Subprogram Call without Using M98/M198

M98 is the usual method of calling a subprogram. (M198 calls a subprogram stored in an external input/output device, such as *Fanuc handy file*, connected via the RS-232C port of the machine. Set parameter 0020 to 0, and parameter 0102 to 3, to enable such a subprogram call. If parameter 0020 is set to 4, M198 calls a subprogram stored in the PCMCIA flash memory card, which is the simplest method on newer control versions, as a PCMCIA slot has now become a standard feature of the control. A different M-code, other than M198, can also be used for subprogram call, by storing the desired number, chosen between 1 and 255, in parameter 6030. If the stored value is 0, M198 is used.) However, if the macro programming feature is enabled on the machine, it is possible to call up to nine subprograms using certain user-defined M-codes. It is also possible to call a subprogram using a T-code. Note that since subprograms are being called, no arguments can be specified for passing on initial data to the local variables of the called program, as in the case of the M98 call. All the rules, which apply to the M98 call, apply to these call methods also.

Subprogram Call Using an M-Code

Any number between 1 and 99,999,999 can be selected as an M-code number to call a subprogram. The maximum permissible nine numbers are stored in parameters 6071 through 6079. The M-codes defined by these parameters call subprograms O9001 through O9009. The correspondence between the program numbers and the parameter numbers

Program Number	Parameter Number
09001	6071
09002	6072
09003	6073
09004	6074
09005	6075
09006	6076
09007	6077
09008	6078
09009	6079

TABLE 7.3   Correspondence between Program Numbers and Parameter Numbers, for Subprogram Call Using an M-Code

is given in Table 7.3. As an example, if the subprogram O9001 is desired to be called by M100, parameter 6071 would need to store 100. Thereafter, M100 would become equivalent to M98 P9001. Though no arguments are allowed, the called subprogram can be repeatedly executed up to 9999 times, using an L-word. For example, M100 L5, which is equivalent to M98 P9001 L5, calls subprogram O9001 five times in succession.

There is a restriction that this method of subprogram call cannot be used in programs called by G-codes (other than G65/G66), M-codes (other than M98/M198), or T-codes. In fact, in general, **the methods of calling a macro/subprogram, using G-code/M-code/T-code, cannot be used in programs called by any of these methods.** The G-code/M-code/T-code used in these programs are treated as **ordinary codes with the original function.** For example, if parameter 6071 contains 3, M03 will call subprogram O9001. But, if M03 is again used in O9001, it would be treated as spindle start command. This "restriction" can be very advantageously used, as explained in the following example.

If the spindle is rotating, the command to change the direction of rotation abruptly (counterclockwise to clockwise or vice versa) may cause excessive load on the spindle motor, due to inertia. Therefore, it may be desirable to insert M05, with a 5-second dwell, while switching between M03/M04. By storing 3 in parameter 6071, and defining O9001 in the given manner, M03 used in any program (the main program, a subprogram called by M98/M198, or a macro called by G65/G66, that is, a program not called by G-code/M-code/T-code) will first stop the spindle, and dwell for 5 seconds, before starting clockwise rotation. Thus, M03 gets modified to have a similar but extended function.

O9001;   (The program number corresponding to parameter number 6071, as per Table 7.3)

M05;   (Spindle stop)

G04 X5;   (Dwell for 5 seconds, to allow the spindle to stop completely)

M03;   (This M-code is an **ordinary** M-code, with its predefined meaning. So, clockwise rotation would start)

M99;   (Return to the calling program)

M04 also can be modified in a similar manner, by storing 4 in parameter 6072, and defining O9002 in the given manner.

O9002;   (The program number corresponding to parameter number 6072, as per Table 7.3)

M05;   (Spindle stop)

G04 X5;   (Dwell for 5 seconds, to allow the spindle to stop completely)

M04;   (This M-code is an ordinary M-code, with its predefined meaning. So, counter-clockwise rotation would start)

M99;   (Return to the calling program)

The only drawback of the modified M03/M04 code is that even when the spindle is stationary, that causes a dwell of 5 seconds. However, in comparison with the total machining time of a job, this much delay can be ignored.

Subprogram Call Using a T-Code

This method is not permitted, unless parameter 6001#5 is set to 1. Using this method, **only one** subprogram O9000 can be called. Any number between 1 and 99,999,999 can be selected for the T-code (i.e., T1 through T99999999 are permitted). After the execution of the T-code, the specified T-code number automatically gets stored in the common variable #149 (which may or may not be used). For example, if T20 is used to call the subprogram, #149 would store 20.

There are several limitations of this method. First, only one subprogram (O9000) can be called. Secondly, no parameters can be specified in the T-code block (since it is a subprogram call). However, up to 9999 repetitions are possible through an L-word. Note that this method **disables the tool change function** of the T-code in the main program, in a subprogram called by M98/M198, and in a macro called by G65/G66. For example, T0101 (which is same as T101) will not change the tool (with or without M06); it will simply call O9000, as a subprogram. (Irrespective of the number used in the T-code, only O9000 is called. If O9000 is not defined, the "NUMBER NOT FOUND" alarm would occur.) However, as already stated, in a program called by G-code (other than G65/G66), M-code (other than M98/M198), or T-code, the T-code is treated as an **ordinary code with the usual tool change function**. This feature can be used to modify the tool change function of a T-code, as explained in the following text.

It is dangerous to change the tool unless the turret is at the home position, because of a possible interference between the rotating tools and the chuck/workpiece/tailstock. Therefore, G28 U0 W0 command must be given before issuing a tool change command. If a programmer forgets to do this, it may cause an accident on the machine. It is, however, possible to redefine the tool change command, to send the turret to the home position automatically, before changing the tool. First set parameter 6001#5 to 1, and then define O9000 in the given manner. Then, T0707 (say) in the calling program will bring tool number 7 to the cutting position, choosing offset number 7, **after** sending the turret to the home position.

O9000;	(The permitted program number for calling a subprogram, using a T-code)
G28 U0 W0;	(Home position return)
T#149;	(This is the usual tool change command, with offset number. #149 stores the number used with the T-code in the calling program. So, the specified T-code is executed here with the usual tool change function)
M99;	(Return to the calling program)

By now, it should be clear that the purpose of redefining the existing G-codes, M-codes, and T-codes is to incorporate extended functions into them. This feature should never be used to change the basic meaning of any code. Always use an unused number for defining codes with new functions.

As another example, one may try to redefine G01, such that if the specified feedrate is more than 100 mm/min for tool number 7, the feedrate used in linear interpolation would be 100 mm/min only. (This is an arbitrary requirement, for the sake of illustration only.) Such a macro, however, cannot be written unless one fully knows the methods of argument specification. Therefore, this example will be taken up in the next section, with the additional purpose of explaining the use of argument specification.

7.5   Argument Specification

In a macro call, certain letter addresses can be used for passing on the desired initial values to the local variables of the called macro. This is called argument specification. Two methods of argument specification are available:

- Argument specification I, which uses all the letters of the alphabet, except L, O, N, G, and P, once each.

- Argument specification II, which uses A, B, and C once each, and also uses I, J, and K up to 10 times each.

The type of argument specification is determined automatically according to the letters used.

Argument Specification I

This type of argument specification can define variables #1 through #26, except #10, #12, #14, #15, and #16. Variables #27 through #33 cannot be defined. The correspondence between the letter addresses and the variable numbers is given in Table 7.4. The prohibited letter addresses can be easily remembered by the phrase "LONG Program."

As an example of argument specification I, G65 P8000 L2 A10 D20 H30 I40 M50 Q60 Z70 would call program number O8000, as a macro, with its local variables initially set as #1 = 10, #7 = 20, #11 = 30, #4 = 40, #13 = 50, #17 = 60, #26 = 70, and the remaining local variables set to null. The word L2 would cause two successive executions of the macro, before returning to the calling program. However, as already discussed in Sec. 7.3, in the second execution (as well as in all the subsequent executions, if the L-word has a value greater than 2), the local variables do not use the specified values in the G65 block as their initial values. Instead, the updated values in the first (previous) execution are used as initial values in the second (subsequent) execution.

Since the CNC uses the word address format, the arguments of G65/G66, including P-word and L-word, can be specified in any order (but no argument should appear to the left of G65/G66). However, I, J, and K, if used, must be specified **alphabetically**. Hence, for example, ... K_ I_ ... must be replaced by ... I_ K_ ..., otherwise it

Address	Variable Number	Address	Variable Number
A	#1	Q	#17
B	#2	R	#18
C	#3	S	#19
D	#7	T	#20
E	#8	U	#21
F	#9	V	#22
H	#11	W	#23
I	#4	X	#24
J	#5	Y	#25
K	#6	Z	#26
M	#13		

TABLE 7.4   Correspondence between Letter Addresses and Variable Numbers in Argument Specification I

would be interpreted in argument specification II, changing the meaning completely.

Argument Specification II

Argument specification II uses A, B, and C once each. Apart from these three letters, it uses I, J, and K up to 10 times each (i.e., 10 independent sets of I, J, and K are used). Thus, it can pass initial values to all the 33 local variables. The correspondence between the letter addresses and the local variables is given in Table 7.5. Note that I_1 through I_{10}, J_1 through J_{10}, and K_1 through K_{10} are written, respectively, as I, J, and K only in the macro-calling block. We use the subscripts (which indicate the set number: set 1 through set 10) for our reference only. The control identifies the subscript (set number) according to the sequence in which the letter addresses appear. For example, if I appears 4 times in the argument list, the first occurrence means I_1, the second occurrence means I_2, the third occurrence means I_3, and the

Address	Variable Number	Address	Variable Number
A	#1	I_6	#19
B	#2	J_6	#20
C	#3	K_6	#21
I_1	#4	I_7	#22
J_1	#5	J_7	#23
K_1	#6	K_7	#24
I_2	#7	I_8	#25
J_2	#8	J_8	#26
K_2	#9	K_8	#27
I_3	#10	I_9	#28
J_3	#11	J_9	#29
K_3	#12	K_9	#30
I_4	#13	I_{10}	#31
J_4	#14	J_{10}	#32
K_4	#15	K_{10}	#33
I_5	#16		
J_5	#17		
K_5	#18		

TABLE 7.5 Correspondence between Letter Addresses and Variable Numbers in Argument Specification II

fourth occurrence means I_4, referring to variables #4, #7, #10, and #13, respectively.

I, J, and K, corresponding to a particular subscript, always appear in the **same** sequence. In other words, these three always appear as an **ordered set**. All three entries of a set need not be defined. For example, I_ J_ I_ K_ K_ J_ means I_1_ J_1_ I_2_ K_2_ K_3_ J_4_ where K_1, J_2, I_3, J_3, I_4, and K_4 are not defined. If we use brackets to separate the sets, it would appear as (I_ J_) (I_ K_) (K_) (J_), where the brackets indicate different sets, the set number increasing from left to right, starting from 1. Note that the J next to K_3 has to be J_4 because K_3 marks the end of the third set, after which the fourth set starts.

The advantage of argument specification II is that all the 33 local variables can be assigned initial values. In addition, since I, J, and K appear in a group, this method of argument specification is very convenient for passing data such as three-dimensional coordinates. However, argument specification I is more commonly used because of its simplicity and easy interpretation of letter addresses. For example, if data for depth of a hole, spindle speed, and feedrate are to be passed on, D, S, and F letters can be used in argument specification I, whereas any three among A, B, C, I, J, and K would have to be used in argument specification II. Obviously, the first type is preferable because it is easy to recall which letter address refers to what.

Mixed Argument Specification

Mixture of argument specification I and argument specification II is permitted, though it is not advisable because it may cause confusion. Note that if only A, B, C, I, J, and K are used, these refer to variables #1 through #6 in **both** the argument specification methods. When a letter other than these six appears, it is interpreted in argument specification I. On the other hand, if any letter among I, J, and K gets repeated (or a different order is used, such as I, K, and J), it is interpreted in argument specification II. If, by chance, the same variable is referred to by two letters (using the two methods), the one specified **later** in the argument list is used to pass the value.

As an example, G65 P8000 L2 E10 J20 J30 J40 H50 K60 K70 F80, which calls program number O8000 twice, uses mixed argument specification. Therefore, the arguments are interpreted left to right, as indicated in Table 7.6. As a result, variables #8, #5, #11, #12, #15, and #9 are assigned 30, 20, 50, 60, 70, and 80, respectively.

It must always be kept in mind that I, J, and K appear as a set, in the same sequence. If any of these gets repeated, or appears in a different sequence, it belongs to a different set. Therefore, in the given example, the three Js refer to J_1, J_2, and J_3, respectively (left to right), and the K after J_3 would be K_3, because only K_3, I_4, and J_4 can appear next to J_3 (in argument specification II). Following similar logic, verify the following:

	G65 P8000 L2 E10 J20 J30 J40 H50 K60 K70 F80			
Argument	Argument Specification Type	Referred Variable	Assigned Value	Comment
E10	Argument specification I	#8	10	
J20	Both specifications equivalent	#5	20	
J30	Argument specification II (J$_2$)	#8	30	Overwrites old value
J40	Argument specification II (J$_3$)	#11	40	
H50	Argument specification I	#11	50	Overwrites old value
K60	Argument specification II (K$_3$)	#12	60	
K70	Argument specification II (K$_4$)	#15	70	
F80	Argument specification I	#9	80	

TABLE 7.6 Example of Mixed-Type Argument Specification

1. Argument specification: E10 K60 J20 J30 J40 H50 K60 K70 F80

   Variable assignment: #6 = 60, #8 = 20, #9 = 80, #11 = 50, #14 = 40, #15 = 60, #18 = 70

2. Argument specification: B20 A10 D40 J50 K60

   Variable assignment: #1 = 10, #2 = 20, #5 = 50, #6 = 60, #7 = 40

3. Argument specification: B20 A10 D40 K60 J50

   Variable assignment: #1 = 10, #2 = 20, #6 = 60, #7 = 40, #8 = 50

4. Argument specification: I10 I10 I10 I10 D40 K60 J50

   Variable assignment: #4 = 10, #7 = 40, #10 = 10, #13 = 10, #15 = 60, #17 = 50

An Example of a G-Code Macro Call with Arguments

Having discussed the methods of argument specification, we are now in a position to write a macro for redefining G01, a problem that was introduced at the end of Sec. 7.4. For redefining G01, one would need to store 1 in parameter number 6051 (or in some other parameter in the range 6050 to 6059). The macro corresponding to parameter 6051 is O9011 (refer to Table 7.1). Therefore, whenever G01 (or G1) is commanded in a program (main program, a macro called by G65/G66, or a subprogram called by M98/M198), O9011 would be called as a macro. Now, the next task is to write the macro O9011.

A general format for G01, for calling the macro, would be used: G01 X_ Y_ Z_ F_ S_, where the letter addresses would have their usual meanings. Since argument specification I has been used, the macro call would initialize local variables #24, #25, #26, #9, and #19, corresponding to the used letter addresses (X, Y, Z, F, and S, respectively).

Refer to Table 3.10(b) for the meanings of the system variables used in program O9011. To keep things simple, it would be assumed that the macro is being called in the constant rpm mode (G97). The constant surface speed mode (G96) is rarely used on a milling machine, though it is very useful and often used on a lathe.

Though any local and/or common variable can be used to store the results of intermediate calculations, it is better to use common variables only, because local variables are used for passing data to the macro. This will reduce the possibility of programming errors. Moreover, a subsequent modification in the macro, to include additional data to be passed on to the macro, would be easy, as any local variable (which has not been used previously to pass data) can be made use of for storing the additional data. Otherwise, the total number of available common variables is also much more than the number of local variables. Finally, if the result of a calculation needs to be stored permanently (i.e., to be retained even after system reset or a power cycle), it would need to be stored in a permanent common variable. This methodology will be used in all the programs that follow from now onward.

Before trying to write the macro, it is recommended to first write down the step-by-step procedure in plain English. This would simplify program coding to a great extent. Skipping this step is like writing equilibrium equations without making free-body diagrams in an engineering mechanics problem! In fact, if the macro involves a complex logic, a proper flowchart of the algorithm should be made, and its execution trace be analyzed. Unless one separates the logic from the language, macro programming will appear very difficult, which it is not. This approach was recommended in Chap. 5 also. It is being reemphasized here because it is very important.

The algorithm, which has been used in macro O9011, is given below:

1. Find out whether G94 or G95 is active at the time of calling the macro.

2. Find out whether G20 or G21 is active at the time of calling the macro.

3. Find out the feedrate active in the block, immediately preceding the macro calling block.

4. If the G01 block does not contain an F-word, use the feedrate obtained in the previous step.

5. Find out the rpm at the time of calling the macro (i.e., in the immediately preceding block).

6. If the G01 block does not contain an S-word, use the rpm obtained in the previous step.

7. Find out the tool number.

8. If the tool number is not 7, go to step 13.

9. Specify the maximum permissible feedrate (F_{max}) in millimeters per minute.

10. Convert F_{max} to be in millimeters per revolution, if G95 is active, using the formula

$$\text{feed per minute} = \text{feed per revolution} \times \text{rpm}$$

11. Convert F_{max} to be in inches per minute or inches per revolution, if G20 is active, by dividing it by 25.4.

12. If the specified/previous feedrate is greater than F_{max}, use F_{max} as the feedrate.

13. Execute linear interpolation with specified/previous values of X, Y, Z, and S, and appropriate value of F.

14. Return to the calling program.

Look into macro O9011 only after clearly understanding the logic used in the given algorithm. Thereafter, it is only a matter of converting the algorithm into codes.

Recall that a G-code inside a macro, which is called by a G-code (whether same or a different code), is treated as a **standard** G-code, with its predefined meaning. Therefore, in step 13 of the macro called by G01, G01 is interpreted as linear interpolation. In addition, if a particular letter address has a null value, that word is ignored in execution. For example, G00 X10 Y#1 is equivalent to G00 X10, if #1 is null. Therefore, in step 13, using G01 with X, Y, and Z addresses will not cause any problem, even if a value is not assigned for some axis. The general format for calling this macro is G01 X_ Y_ Z_ F_ S_, where not all the letter addresses need to be used.

```
O9011 (REDEFINES G01 ON MILLING MACHINE);
#100 = #4005;              (Stores 94 or 95, corresponding to G94
                            and G95, respectively, whichever was
                            active at the time of calling the macro.
                            This information is needed because one
```

must know whether the specified feedrate in the macro-calling block is in feed per minute or feed per revolution. **Step 1 complete)**

`#101 = #4006;`

(Stores 20 or 21, corresponding to G20 and G21, respectively, whichever was active at the time of calling the macro. This information is needed because one must know whether the specified distances are in inches or millimeters. **Step 2 complete)**

`#102 = #4109;`

(Stores the active feedrate up to the previous block. This information would be needed if G01 is commanded without an F-word. In such a case, the last specified feedrate would be used. **Step 3 complete)**

`IF [#9 EQ #0] THEN #9 = #102;`

(The last specified feedrate being used, in the absence of an F-word in the G01 block. **Step 4 complete)**

`#103 = #4119;`

(Stores the spindle rpm at the time of calling the macro. This information would be needed if G01 is commanded without an S-word. In such a case, the current rpm would be used. Note that this macro has been designed to be called in the constant rpm mode, i.e., when G97 is active. In G96 mode, certain changes would need to be done. However, G96 is rarely used on milling machines. **Step 5 complete)**

`IF [#19 EQ #0] THEN #19 = #103;`

(The current spindle rpm being used, in the absence of an S-word in the G01 block. **Step 6 complete)**

`#104 = #4120;`

(Stores the current tool number, to check if it is tool number 7. **Step 7 complete)**

`IF [#104 NE 7] GOTO 10;`

(If the current tool is other than tool number 7, jump to sequence number N10, to execute G01 without changing the feedrate. **Step 8 complete)**

`#105 = 100;`

(Specify the maximum permissible feedrate, in millimeters per minute, for tool number 7. **Step 9 complete)**

`IF [#100 EQ 95] THEN #105 = #105 / #19;`

(Maximum permissible feedrate converted to be in feed per revolution, for use in G95 mode. **Step 10 complete)**

`IF [#101 EQ 20] THEN #105 = #105 / 25.4;`

(Converts the maximum permissible feedrate to be in inches per minute or inches per revolution, if the program is in G20 mode. **Step 11 complete)**

```
IF [#9 GT #105] THEN #9 = #105;
```
(Specified/previous feedrate clamped to the maximum permissible value. **Step 12 complete**)

```
N10 G01 X#24 Y#25 Z#26 F#9 S#19;
```
(Linear interpolation with specified/ previous values of X, Y, Z, and S, and appropriate value of F. **Step 13 complete**)

```
M99;
```
(Return to the calling program. **Step 14 complete**)

Program O9011 given here is for milling machines. For a lathe, it would need to be modified to delete the Y-word. Moreover, apart from X and Z, U and W are also used on a lathe. Hence, the program for a lathe would need to be called by G01 X_ Z_ U_ W_ F_ S_. One can store 1 in parameter 6052 (say), and define the corresponding program O9012, for redefining G01. Refer to Table 3.10(a) and Table 3.11 for information about the system variables used in this program.

The algorithm is similar to that used in program O9011, except that O9012 takes care of constant surface speed (G96) mode also, for calculating the rpm that is needed in the formula relating feed per minute to feed per revolution, for use in G99 mode. Note, however, that the feedrate would be clamped (if needed) only in the beginning of the linear interpolation move. If feed per revolution (G99) mode is also active in G96 mode, the feedrate (feed per minute) would increase if the cutting diameter decreases while executing linear interpolation (e.g., in facing or taper turning). The given macro has no control over such a varying feedrate. It will only check the feedrate in the **beginning** of the move.

The rpm (N) is related to the constant surface speed (CSS) and the cutting diameter (D), which is same as the current X-position of the tool, by the following formula:

$$N = \frac{1000 \times CSS}{\pi \times D}$$

where CSS is in meters per minute and D is in millimeters and

$$N = \frac{12 \times CSS}{\pi \times D}$$

where CSS is in feet per minute and D is in inches.

The algorithm used in macro O9012 is given below, followed by the program:

1. Find out whether G98 or G99 is active at the time of calling the macro.

2. Find out whether G20 or G21 is active at the time of calling the macro.

3. Find out the feedrate active in the block, immediately preceding the macro calling block.

4. If the G01 block does not contain an F-word, use the feedrate obtained in the previous step.

5. Find out the S-word value active in the immediately preceding block.

6. If the G01 block does not contain an S-word, use the value obtained in the previous step.

7. Find out the tool number, using FIX[T-code value/100], which extracts the left-most two digits from the four-digit (can be a three-digit number also) T-code value. For example, 0102 (corresponding to T0102) would give 01, as the current tool number.

8. If the tool number is not 7, go to step 15.

9. Find out whether G96 or G97 is active at the time of calling the macro.

10. Find out the rpm at the time of calling the macro (i.e., in the immediately preceding block). In G97 mode, it would be equal to the S-word value. In G96 mode, calculation using the given formula would be needed.

11. Specify the maximum permissible feedrate (F_{max}) in millimeters per minute.

12. Convert F_{max} to be in millimeters per revolution if G99 is active, using the formula

feed per minute = feed per revolution × rpm

13. Convert F_{max} to be in inches per minute or inches per revolution, if G20 is active, by dividing it by 25.4.

14. If the specified/previous feedrate is greater than F_{max}, use F_{max} as feedrate.

15. Execute linear interpolation with specified/previous values of X, Z, U, W, and S, and appropriate value of F.

16. Return to the calling program.

```
O9012 (REDEFINES G01 ON LATHE);
#100 = #4005;
```
(Stores 98 or 99, corresponding to G98 and G99, respectively, whichever was active at the time of calling the macro. This information is needed because one must know whether the specified feedrate in the macro calling block is in feed per minute or feed per revolution. **Step 1 complete**)

#101 = #4006;   (Stores 20 or 21, corresponding to G20 and G21, respectively, whichever was active at the time of calling the macro. This information is needed because one must know whether the specified distances are in inches or millimeters. **Step 2 complete**)

#102 = #4109;   (Stores the active feedrate up to the previous block. This information would be needed if G01 is commanded without an F-word. In such a case, the last specified feed-rate would be used. **Step 3 complete**)

IF [#9 EQ #0] THEN #9 = #102;   (The last specified feedrate being used, in the absence of an F-word in the G01 block. **Step 4 complete**)

#103 = #4119;   (Stores the spindle rpm/CSS at the time of calling the macro. This information would be needed if G01 is commanded without an S-word. In such a case, the current rpm/CSS would be used. Note that when G97 is active, the S-word denotes rpm. In G96 mode, it represents CSS. **Step 5 complete**)

IF [#19 EQ #0] THEN #19 = #103;   (The current spindle rpm/CSS being used, in the absence of an S-word in the G01 block. **Step 6 complete**)

#104 = #4120;   (Stores the four-digit current tool code, to check if it is tool number 7)

#104 = FIX [#104 / 100];   (Extracts the tool number from the four-digit tool code. **Step 7 complete**)

IF [#104 NE 7] GOTO 10;   (If the current tool is other than tool number 7, jump to sequence number N10 to execute G01 without changing the feedrate. **Step 8 complete**)

#106 = #4002;   (Stores 96 or 97 depending on which one of G96 and G97 is currently active. **Step 9 complete**)

IF [#106 EQ 97]THEN #107 = #19;   (In G97 mode, the S-word contains rpm that gets stored in #107)

#108 = #5041;   (Stores the current tool position along the X-axis. This equals the diameter being turned **currently**, i.e., at the time of calling G01. Note that the turning diameter continuously changes in taper turning or facing. However, in the formula for calculating the rpm, in G96 mode, the initial diameter, as obtained here, is used. As a result, even if the rpm continuously changes in G96 mode while executing G01, only the **initial** rpm is calculated)

```
#109 = [[1000 * #19] / [3.14159 * #108]];
```
                  (Intermediate calculation. Use of outer brackets ensures that the slash is always interpreted as the division operator, never as the mid-block skip symbol)

```
IF [[#106 EQ 96] AND [#101 EQ 21]] THEN #107 = #109;
```
                  (#107 stores the rpm calculated in CSS and millimeter modes)

```
#110 = [[12 * #19] / [3.14159 * #108]];
```
                  (Intermediate calculation. Use of outer brackets ensures that the slash is always interpreted as the division operator, never as the mid-block skip symbol)

```
IF [[#106 EQ 96] AND [#101 EQ 20]] THEN #107 = #110;
```
                  (#107 stores the rpm calculated in CSS and inch modes. **Step 10 complete**)

```
#105 = 100;
```
                  (Specify the maximum permissible feedrate, in millimeter per minute, for tool number 7. **Step 11 complete**)

```
IF [#100 EQ 99] THEN #105 = #105 / #107;
```
                  (Maximum permissible feedrate converted to be in feed per revolution, for use in G99 mode. **Step 12 complete**)

```
IF [#101 EQ 20] THEN #105 = #105 / 25.4;
```
                  (Converts the maximum permissible feedrate to be in inches per minute or inches per revolution, if the program is in G20 mode. **Step 13 complete**)

```
IF [#9 GT #105]THEN #9 = #105;
```
                  (Specified/previous feedrate clamped to the maximum permissible value. **Step 14 complete**)

```
N10 G01 X#24 Z#26 U#21 W#23 F#9 S#19;
```
                  (Linear interpolation with specified/previous values of X/U, Z/W, and S, and appropriate value of F. **Step 15 complete**)

```
M99;
```
                  (Return to the calling program. **Step 16 complete**)

The purpose of selecting this example was not just to write a macro for a specific application. (In fact, one may never need to redefine G01 on a machine.) This example makes it amply clear that a complex macro such as this, perhaps, cannot be correctly written without first writing the algorithm. However, once the algorithm is ready, what remains is the mechanical job of converting it into codes. Without adopting this approach, a new learner would, perhaps, never be able to master macro programming; even a seasoned programmer is likely to make mistakes.

A few more things remain to be discussed, though these are not directly related to the topic of this chapter. They are, however, discussed here because these do not merit a separate chapter, and this is the last chapter dealing with the basics of macro programming.

7.6   Processing of Macro Statements

It is first necessary to understand the concept of *buffering* during program execution, before discussing how macro statements are processed by the control.

What Is Buffering?

Though the execution of a program is block by block, the control pre-reads the NC statement to be executed next, for smooth machining. The next NC statement is read and interpreted in advance, but executed only after the execution of the previous statement is complete. If both are movement commands, the execution of the second command starts a little before the end point of the first command is reached, which results in slightly rounded corners. This small distance is called *in-position width*, which is specified in parameter 1827. For obtaining sharp corners, specify a small value in this parameter. Alternatively, on a milling machine, run the program in the *exact stop mode*, G61. However, rounding the corners speeds up the execution, due to less deceleration/acceleration effect, and, in most of the cases, very sharp corners are not needed. In radius compensation mode, two NC statements are read in advance, and if one of these is not a movement command in the plane of compensation, then one more NC statement is preread. Such a prereading operation is referred to as buffering.

How Many Blocks Are Buffered?

Up to three NC statements are buffered, depending on specific situations, though the blocks next to G31, M00, M01, M02, M30, and M-codes specified in parameters 3411 to 3420 are not buffered. However, the control tries to buffer as many macro statements as it can, during the execution of the current NC statement.

When Are the Buffered Blocks Processed?

As already mentioned, a buffered NC statement is processed (executed) after the execution of the current NC statement ends (or "nearly" ends). However, a macro statement is **immediately** processed, as soon as it is buffered.

Processing of macro statements depends on how buffering is done by the control. There are four possible situations, in which buffering is done differently. Processing of macro statements in these cases is described next, with the help of examples.

Processing When the Next Block Is Not Buffered

While executing certain M-codes and G31, the control does not buffer the next block. Therefore, if the next block contains a macro statement, it is buffered and processed only after the completion of the processing of the current block:

```
...
N100 G31 X100;   (The next block is not buffered because of G31)
N101 #100 = 1;   (#100 is set to 1, after the execution of G31 is complete)
...
```

Processing in Radius Compensation Cancel (G40) Mode

All the subsequent macro statements, up to the next NC statement, are read and immediately processed:

```
...
N100 G01 X100 F100;   (While executing this block, buffering is done
                       up to N104)
N101 #100 = 1;        (#100 is set to 1 while N100 is being executed)
N102 #101 = 2;        (#101 is set to 2 while N100 is being executed)
N103 #102 = 3;        (#102 is set to 3 while N100 is being executed)
N104 Z100;            (The next NC statement, after N100 block)
N105 #103 = 4;        (This macro statement is not buffered while
                       N100 is being executed)
...
```

Processing in Radius Compensation Mode (Case 1)

For processing in radius compensation (G41/G42) mode, when the next two NC statements are movement commands, in the plane of compensation, all the macro statements, up to the second NC statement after the currently executing NC statement, are read and immediately processed. (This example pertains to a milling machine, assuming that the plane of compensation is G17, i.e., the XY-plane.)

```
...
N100 G01 X100 F100;   (While executing this block, buffering is done
                       up to N106)
N101 #100 = 1;        (#100 is set to 1 while N100 is being executed)
N102 #101 = 2;        (#101 is set to 2 while N100 is being executed)
N103 Y100;            (The first NC statement, after N100 block)
N104 #102 = 3;        (#102 is set to 3 while N100 is being executed)
N105 #103 = 4;        (#103 is set to 4 while N100 is being executed)
N106 X200;            (The second NC statement, after N100 block)
N107 #104 = 5;        (This macro statement is not buffered while
                       N100 is being executed)
...
```

Processing in Radius Compensation Mode (Case 2)

For processing in radius compensation (G41/G42) mode, when one (either one) of the next two NC statements is not a movement command, in the plane of compensation, all the macro statements, up to the third NC statement after the currently executing NC statement, are read and immediately processed. (This example pertains to a milling machine, assuming that the plane of compensation is G17, i.e., the XY-plane.)

```
  . . .
  N100 G01 X100 F100;    (While executing this block, buffering is done
                          up to N109)
  N101 #100 = 1;         (#100 is set to 1 while N100 is being executed)
  N102 #101 = 2;         (#101 is set to 2 while N100 is being executed)
  N103 Y100;             (The first NC statement, after N100 block)
  N104 #102 = 3;         (#102 is set to 3 while N100 is being executed)
  N105 #103 = 4;         (#103 is set to 4 while N100 is being executed)
  N106 Z10;              (The second NC statement, after N100 block.
                          This statement does not cause movement in
                          the plane of compensation)
  N107 #104 = 5;         (#104 is set to 5 while N100 is being executed)
  N108 #105 = 6;         (#105 is set to 6 while N100 is being executed)
  N109 X100;             (The third NC statement, after N100 block)
  N110 #106 = 7;         (This macro statement is not buffered while
                          N100 is being executed)
  . . .
```

Finally, in radius compensation mode, if both the subsequent NC statements, after the currently executing NC statement, do not involve movement in the plane of compensation, this would result in incorrect compensation. The program must be modified to avoid this situation. However, for the sake of discussion, this case is equivalent to the previous case, with regard to buffering of macro statements (for example, if N103 Y100 is replaced by, say, N103 M08, buffering would still be done up to N109 only).

Effect of Buffering of Macro Statements on Program Execution

The purpose of buffering is to speed up the program execution by performing the calculations in advance. Normally, it only improves the performance of the machine, but there are situations when buffering is not desirable, and it must be somehow suppressed. For example, assume that system variable #1100 (which corresponds to signal F54.0 in the PMC ladder diagram) has been used to operate an external device, connected to the output terminal Y5.0 (which can be located on the terminal strip of the input/output module of the PMC), by adding the following rung to the existing ladder diagram:

```
          F54.0                              Y5.0
       |——| |————————————————————————————( )——|
```

This rung would set the output device on or off, depending on the current status (1 or 0) of variable #1100. Now, if the requirement is to switch on the device somewhere in the middle of program execution, wait for 5 seconds, and then switch it off and restart the execution, one may insert the following blocks in the program, at the desired place:

```
#1100 = 1;
G04 X5;
#1100 = 0;
```

Unfortunately, this would not work. The output would indeed be turned on, but while performing the dwell for 5 seconds, the next macro statement (#1100 = 0) would be buffered and immediately executed, switching off the output instantly. Thus, even though the execution would dwell for 5 seconds, the output would switch on only momentarily. Therefore, in order to have the desired effect, it is necessary to suppress buffering in this case. This can be very simply done by inserting a blank NC statement (a semicolon, the EOB symbol, is treated as a blank NC statement) after the G04 block:

```
#1100 = 1;
G04 X5;
;
#1100 = 0;
```

Fanuc control does not define a G-code to limit buffering to the desired number of blocks, unlike Haas control that has G103 for this purpose. However, as mentioned earlier, it is possible to define certain M-codes (in the range 0 to 255), for the sole purpose of preventing buffering of the following block. For example, if 100 is stored in any parameter in the range 3411 to 3420, M100 would prevent buffering of the following block. Then, instead of inserting a blank NC statement, in the previous example, M100 can be commanded. This method is better because a different programmer may consider the extra semicolon unnecessary, and may even choose to delete it! Note, however, that buffering should not be suppressed in radius compensation mode, because otherwise, the compensation would not be properly implemented by the control, resulting in incorrect machining.

CHAPTER 8

Complex Motion Generation

8.1 Introduction

So far, only the basic tools of macro programming have been discussed. With the help of these, it is possible to generate complex toolpaths, apart from several other applications that are discussed in subsequent chapters. Some examples of complex motion generation are given in this chapter. Generally speaking, it is easier to do it on a milling machine (in 2D machining), than on a lathe, because the generated toolpath would have to be used in conjunction with a suitable canned cycle (G71/G72/G73) on a lathe, to avoid large depth of cut. Therefore, milling examples are discussed first.

8.2 Arc with Uniformly Varying Radius

Consider the slot shown in Fig. 8.1, which is in the form of an arc (AB), with uniformly increasing radius. The start and the end radii are 35 mm (CA) and 40 mm (CB), respectively. The angle of arc is 60°, the start angle being 15°. The workpiece zero point (origin or datum of the chosen WCS, among G54, G55, . . ., G59) is at O. The center of the arc (C) is located at coordinates (30, 20). The depth of the slot is assumed to be 3 mm.

Since the radius of the arc is continuously varying, circular interpolation cannot be used in this case. The only way is to locate closely spaced points on the arc AB, and join these by linear interpolation. The spacing between these points has to be carefully selected. A small value would improve the smoothness of the arc. However, too small a value would increase the calculations, which may result in sluggish tool movement. (Another and the main cause of the sluggish motion is that acceleration/deceleration is involved in every toolpath segment. Therefore, the effective feedrate might be much smaller than the specified feedrate.)

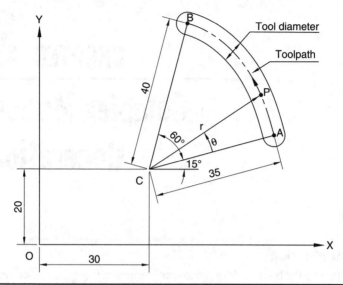

Figure 8.1   Arc with uniformly increasing radius.

The coordinates of some point P on the arc (X_P, Y_P), as a function of angle θ, can be found using the following equations, assuming the radius r changes with θ in a uniform (linear) manner:

$$r = r_A + \frac{(r_B - r_A)\theta}{\angle ACB}$$

$$X_P = X_C + r\cos(\theta + \theta_A)$$

$$Y_P = Y_C + r\sin(\theta + \theta_A)$$

where r_A = CA (start radius)
r_B = CB (end radius)
$\angle ACB$ = included angle of the arc
θ_A = angle of CA with the X-axis (start angle)
X_C = X-coordinate of the arc center C
Y_C = Y-coordinate of the arc center C

At the start point of the arc (A), the value of θ is zero. Thereafter, it gradually increases to reach the end point (B). This forms the basis of an algorithm for this problem.

The first step in developing an algorithm for the macro is to decide how to use the macro. Here, it is assumed that the tool is 2 mm above the start point of the arc (A) at the time of calling the macro, and it comes back to the same Z-level at the end point of the arc (B) when execution of the macro ends. It is also assumed that the calling program uses millimeters, (G21), the XY-plane (G17), feedrate in millimeters per minute (G94), absolute coordinates (G90), no radius compensation

(G40), and no active canned cycle (G80). It is further assumed that the correct tool (slotdrill of 6 mm diameter) is held by the spindle, and it is rotating with the correct rpm. Finally, the choice of WCS (say, G54) and tool length compensation (G43 H_) are assumed to be correct. With these assumptions, the macro cannot be said to be quite general, but the purpose here is to describe the technique of generating an arc with varying radius. The arc, of course, would be defined in terms of a number of variable parameters, making it quite general. Once the basic program is ready, it becomes easy to make it more general, as per requirement. In fact, it is not recommended to consider all the possibilities in the very beginning, as such an approach makes programming difficult, leaving enough scope for logical errors.

The parameters of the slot, the selected letter addresses for these, and the associated local variables, as per argument specification I (refer to Table 7.4), are given in Table 8.1. Since these letter addresses would be used for passing data in the G65/G66 macro call, their choice should reflect their actual physical meanings, as much as possible. This would help in recalling which letter address refers to which parameter. In the present case, the first letters of the parameters have been chosen as letter addresses, except A and F, where it would not be possible.

Now, the algorithm for this problem, suitable for WHILE_DO_ END loop, can be written (note that the specified condition is checked in the beginning of the loop, in a WHILE statement):

1. Make a plunge entry up to the specified depth, with the specified plunge feedrate.
2. Choose the increment in angle, $\Delta\theta$, and assign $\theta = \Delta\theta$.
3. If θ is greater than the included angle of the arc, go to step 8.

Parameters of the Slot	Selected Letter Addresses	Associated Variables
Start angle	A	#1
Included angle of the arc	I	#4
Start radius	S	#19
End radius	E	#8
Depth of slot	D	#7
X-coordinate of the arc center	X	#24
Y-coordinate of the arc center	Y	#25
Plunge feedrate	F	#9
Milling feedrate	M	#13

TABLE 8.1 Parameters of the Slot of Fig. 8.1 and the Selected Letter Addresses

4. Calculate r, X_p, and Y_p, using the given formula.

5. Move to the calculated coordinates (X_p, Y_p) by G01, with the specified milling feedrate.

6. Increment angle ($\theta = \theta + \Delta\theta$).

7. Go to step 3.

8. Retract the tool to 2 mm above the workpiece.

9. Return to the calling program.

Program number 8014 is based on this algorithm. For making the slot of Fig. 8.1, this macro would need to be called by

```
G65 P8014 A15 I60 S35 E40 D3 X30 Y20 F20 M60;
```

where the plunge feedrate and milling feedrate are taken as 20 and 60 mm/min, respectively.

`O8014 (ARC WITH VARYING RADIUS);`	
`G01 Z-[ABS[#7]] F#9;`	(Plunge entry with the specified plunge feedrate at the start point A. Use of ABS function accepts both positive and negative values for the depth of the slot. **Step 1 complete**)
`#100 = 0.1;`	($\Delta\theta$ selected, in degrees)
`#101 = #100;`	(θ set equal to $\Delta\theta$. **Step 2 complete**)
`WHILE [#101 LE #4] DO 1;`	(Exit from the loop if θ is greater than the included angle of the arc. **Step 3 complete**)
`#102 = #19 + [[#8 - #19]* #101 / #4];`	(r calculated)
`#103 = #24 + #102 * COS[#101 + #1];`	(X_p calculated)
`#104 = #25 + #102 * SIN[#101 + #1];`	(Y_p calculated. **Step 4 complete**)
`X#103 Y#104 F#13;`	(Move to (X_p, Y_p) with the specified milling feedrate. **Step 5 complete**)
`#101 = #101 + #100;`	(Angle θ incremented by the chosen step. **Step 6 complete**)
`END 1;`	(End of WHILE loop. **Step 7 complete**)
`G00 Z2;`	(The tool is now at the end point B of the arc. Here, it is retracted to 2 mm above the workpiece. **Step 8 complete**)
`M99;`	(Return to the calling program. **Step 9 complete**)

This macro would work for both increasing and decreasing radii (i.e., also when $r_A > r_B$). However, for a clockwise arc (i.e., when positions of A and B get interchanged, and machining is still desired from A to B, in Fig. 8.1), the following changes would be needed:

- Specify a negative value for $\Delta\theta$ (e.g., #100 = –0.1).

- In the formula for r, the absolute value of θ (which becomes negative when $\Delta\theta$ is assigned a negative value) should be used. So, replace #101 by ABS[#101] in the block that calculates r. There would be no change in the blocks that calculate X_p and Y_p. The other way would be to specify a negative value, in the macro-calling block, for the included angle if it measures in the clockwise direction, when moving from OA to OB. This would obviate the need for any change in the program.

- The WHILE block should compare the absolute values.

Modifying the macro for slightly different requirements is not a good practice, because a less experienced programmer may inadvertently spoil the macro developed meticulously by some expert programmer. Moreover, other users of the machine may not be aware of the changes made in the macro. In fact, it is because of this very reason that macros are usually edit-protected.

In view of the foregoing discussion, it would be better to define one more local variable (say, #17 that refers to letter address Q) for passing the value of increment in angle ($\Delta\theta$) to the macro, such that a positive increment value would make a counter-clockwise arc and a negative value would make a clockwise arc. Moreover, since there might be a confusion regarding the sign of the included angle, the macro should accept both signs. An advantage of passing the value of $\Delta\theta$ is that the user of the macro can also control the smoothness of the arc. Program number 8015 is one such macro that can be called by

```
G65 P8015 A15 I60 S35 E40 D3 X30 Y20 F20 M60 Q-0.1;
```

or

```
G65 P8015 A15 I-60 S35 E40 D3 X30 Y20 F20 M60 Q-0.1;
```

for a clockwise arc. The readers may try to make the macro more general to suit specific applications. Once the core of the program is ready, modifying it is not a big deal, as can be seen in the present case where O8014 has been modified to O8015.

```
O8015 (ARC WITH VARYING RAD - MODIFIED);
G01 Z-[ABS[#7]] F#9;
```
(Plunge entry with the specified plunge feedrate at the start point A. Use of ABS function accepts both positive and negative values for the depth of the slot. **Step 1 complete**)

```
#100 = #17;
```
($\Delta\theta$ set to the specified value in the Q-word, which can be both positive and negative, for CCW and CW arcs, respectively)

```
#101 = #100;
```
(θ set equal to $\Delta\theta$. **Step 2 complete**)

```
WHILE [ABS[#101] LE ABS[#4]] DO 1;
```
(Exit from the loop if the magnitude of θ is greater than the magnitude of the included angle of the arc. ABS[#4] would accept both positive and negative values for the included angle. **Step 3 complete**)

```
#102 = #19 + [[#8 - #19] * ABS[#101] / ABS[#4]];
```
(r calculated. ABS[#4] would accept both positive and negative values for the included angle)

```
#103 = #24 + #102 * COS[#101 + #1];
```
(X_p calculated)

```
#104 = #25 + #102 * SIN[#101 + #1];
```
(Y_p calculated. **Step 4 complete**)

```
X#103 Y#104 F#13;
```
(Move to (X_p, Y_p) with the specified milling feedrate. **Step 5 complete**)

```
#101 = #101 + #100;
```
(Angle θ incremented by the specified step value. The angle actually gets decremented if the step value is negative, for a clockwise arc. **Step 6 complete**)

```
END 1;
```
(End of WHILE loop. **Step 7 complete**)

```
G00 Z2;
```
(The tool is now at the end point B of the arc. Here, it is retracted to 2 mm above the workpiece. **Step 8 complete**)

```
M99;
```
(Return to the calling program. **Step 9 complete**)

This macro would make the slot in one pass. If the depth of the slot is large, say, 10 mm, it may not be practical to mill it in one pass. If the maximum permissible depth of cut is, say, 3 mm, four passes would be needed (at, say, 3-, 6-, 9-, and 10-mm depths, respectively). Therefore, the next task is to take care of this requirement. One can make use of the logic used in the flowchart given in Fig. 6.2. The following steps would need to be incorporated in the algorithm given earlier:

1. Move the tool to Z0 position.
2. If the specified/calculated depth is more than 3 mm, then
   - Make an arc of 3-mm depth (incremental distance from the previous Z-level), using the algorithm of the previous program (this step would appear as a nested WHILE loop in this program).
   - After reaching the end point of the arc, retract the tool above the workpiece and bring it back to the start point of the arc, at the same Z-level, with rapid rate (G00).
   - Now bring the tool to the previous Z-level (with rapid rate up to 1 mm above the previous Z-level, followed by feedrate in the last 1 mm travel).

- The slot milling in the previous pass reduces the required depth by 3 mm, so, set depth = depth – 3.
- Loop back to step 2.

3. Make the arc at the specified/calculated depth (incremental distance from the previous Z-level). Another WHILE loop will be needed for this step.

4. Retract the tool to 2 mm above the workpiece.

5. Return to the calling program.

Program number 8016 is based on this algorithm, which uses a two-level nested WHILE statement. It can be called by

```
G65 P8016 A15 I60 S35 E40 D10 X30 Y20 F20 M60 Q-0.1;
```

or

```
G65 P8016 A15 I-60 S35 E40 D10 X30 Y20 F20 M60 Q-0.1;
```

for a clockwise arc of 10-mm depth. The machining would be done in four passes (once each at Z = –3, –6, –9, and –10 mm)

`O8016 (VAR RAD ARC WITH ARBIT DEPTH);`	
`G01 Z0 F#9;`	(Tool made to touch the workpiece at the start point of the arc. **Step 1 complete**)
`#107 = ABS[#7];`	(Positive value of the specified slot depth in the macro calling block would be used further, which permits both positive and negative values in the macro calling block)
`WHILE [#107 GT 3] DO 1;`	(**Step 2** starts here)
`G91 G01 Z-3 F#9;`	(Plunge entry to 3 mm below the current Z-level, at the start point A, with the specified plunge feedrate)
`#100 = #17;`	($\Delta\theta$ set to the specified value in the Q-word, which can be both positive and negative, for CCW and CW arcs, respectively. This block can be placed outside the loop also, i.e., before the WHILE_DO_ block, because #100 is not redefined by the loop)
`#101 = #100;`	(θ set equal to $\Delta\theta$)
`WHILE [ABS[#101] LE ABS[#4]] DO 2;`	
	(Exit from the loop if the magnitude of θ is greater than the magnitude of the included angle of the arc. ABS[#4] would accept both positive and negative values for the included angle)
`#102 = #19 + [[#8 - #19] * ABS[#101] / ABS[#4]];`	
	(r calculated. ABS[#4] would accept both positive and negative values for the included angle)

```
#103 = #24 + #102 * COS[#101 + #1];
```
            (X_p calculated)
```
#104 = #25 + #102 * SIN[#101 + #1];
```
            (Y_p calculated)
```
G90 X#103 Y#104 F#13;
```
            (Move to (X_p, Y_p) with the specified milling feedrate)
```
#101 = #101 + #100;
```
            (Angle θ incremented/decremented by the chosen step)
```
END 2;
```
            (End of WHILE loop. Arc made)
```
#105 = #5043;
```
            (The current absolute Z-coordinate of the tool stored in #105)
```
G00 Z2;
```
            (The tool is now at the end point B of the arc. Here, it is retracted to 2 mm above the workpiece)
```
X[#24 + #19 * COS[#1]] Y[#25 + #19 * SIN[#1]];
```
            (Tool brought back to the start point A)
```
Z[#105 + 1];
```
            (Rapid to 1 mm above the previous Z-level)
```
G01 Z#105 F#9;
```
            (Feed motion in the last 1-mm travel)
```
#107 = #107 - 3;
```
            (Calculates the remaining depth of the slot, to be machined further)
```
END 1;
```
            (Loop back to the start of step 2. **Step 2 complete**)
```
G91 G01 Z-#107 F#9;
```
            (Plunge entry to the final depth, at the start point A, with the specified plunge feedrate)
```
#100 = #17;
```
            (Δθ set to the specified value. This statement is redundant, because #100 is already defined)
```
#101 = #100;
```
            (θ set equal to Δθ)
```
WHILE [ABS[#101] LE ABS[#4]] DO 3;
```
            (Exit from the loop if the magnitude of θ is greater than the magnitude of the included angle of the arc. Since it is not a nested WHILE loop, one may choose to use even 1 or 2, in place of 3, as loop identification number)
```
#102 = #19 + [[#8 - #19] * ABS[#101] / ABS[#4]];
```
            (r calculated)
```
#103 = #24 + #102 * COS[#101 + #1];
```
            (X_p calculated)
```
#104 = #25 + #102 * SIN[#101 + #1];
```
            (Y_p calculated)
```
G90 X#103 Y#104 F#13;
```
            (Move to (X_p, Y_p) with the specified milling feedrate)
```
#101 = #101 + #100;
```
            (Angle θ incremented/decremented by the chosen step)
```
END 3;
```
            (End of WHILE loop. Arc made. **Step 3 complete**)
```
G00 Z2;
```
            (The tool is now at the end point B of the arc. Here, it is retracted to 2 mm above the workpiece. **Step 4** complete)
```
M99;
```
            (Return to the calling program. **Step 5 complete**)

This program is quite general now, except that it assumes certain control conditions. It is left as an exercise for the users to make the macro suitable for any control conditions. The approach would be similar to that used in the macros of Chap. 7 (O9011 and O9012). Moreover, the macro assumes that the maximum permissible depth of cut in a pass is 3 mm. A local variable can be used to supply this value to the macro. This is desirable because different cutting conditions require different cutting parameters. The readers should try to do this also, which is pretty straightforward (just replace 3 by the chosen local variable at required places).

An important point regarding the style of this program must be observed: the local variables, which were used for value assignment through the G65/G66 statement, have not been redefined inside the macro. The temporary calculation results have been stored in common variables. This practice reduces the possibility of errors in the macro that may require using the **original** specified values at several places. So, if the specified value of some local variable is to change, as per the logic of the macro, store it in some common variable, and use it instead of the local variable. The #107 = ABS[#7] statement does exactly the same thing, to take care of reducing the depth of the slot, after each pass. #7 has not been redefined.

8.3 Helical Interpolation with Variable Radius

Helical interpolation is available on three-axis milling machines, as an option. It is, basically, a circular interpolation only, superimposed with **synchronized** linear motion along the third axis, such that the drive motors for all the three axes reach the specified end point at the same time, while moving at uniform rate (though their speeds would be different from one another). The specified feedrate is maintained along the **projected toolpath** on the plane of circular interpolation. Therefore, the feedrate along the linear axis would be (which is automatically calculated by the control)

$$\text{specified feedrate} \times \frac{\text{length of linear axis}}{\text{length of circular arc}}$$

Obviously, the actual feedrate of the tool, that is, feedrate along the toolpath, would be more than the specified feedrate. It would, in fact, be equal to

$$\sqrt{(\text{feedrate along linear axis})^2 + (\text{specified feedrate})^2}$$

On a three-axis milling machine, helical interpolation is possible along the X-, Y-, and Z-axes, corresponding to G19 (YZ-plane), G18 (XZ-plane), and G17 (XY-plane), respectively. Helical interpolation along the Z-axis is shown in Fig. 8.2, which can be commanded as

```
<Place the tool at the start point S>
G17 G03 X_ Y_ Z_ R_ F_; or G17 G03 X_ Y_ Z_ I_ J_ F_;
```

where X_, Y_, Z_ = coordinates of the end point P
  R_ = radius of the helix/circular interpolation
  I_, J_ = X-, Y-coordinates of the axis of the helix with respect to the start point S
  F_ = feedrate along the projected toolpath on the XY-plane

G03 generates a counter-clockwise helix, as in Fig. 8.2. Replace it by G02 for a clockwise helix. The direction of the helix (CW/CCW) is determined by the right-hand rule where the thumb points in the positive direction of the linear axis. Thus, if the toolpath follows the direction of the the curled fingers, it is called a counter-clockwise helix.

Helical interpolations along X- and Y-axes are done in a similar manner. (Note that the chosen orders of X_, Y_, and Z_ are only for our convenience in identifying the plane of associated circular interpolation. The plane of circular interpolation is decided by G17/G18/G19. In the word address format, which the CNC machines use, the order of the arguments of a G-code is not important, barring a few exceptions such as the order of I-, J-, and K-word in the G65/G66 argument list):

```
G18 G02/G03 X_ Z_ Y_ R_ F_; or G18 G02/G03 X_ Z_ Y_ I_ K_ F_;
G19 G02/G03 Y_ Z_ X_ R_ F_; or G19 G02/G03 Y_ Z_ X_ J_ K_ F_;
```

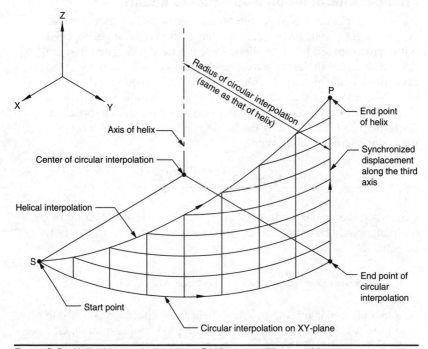

FIGURE 8.2 Helical interpolation along Z-axis on a milling machine.

There are two limitations of the built-in helical interpolation, described above:

- The radius/pitch of the helix cannot be varied.
- The feedrate along the actual toolpath is more than the specified feedrate.

We will now develop a macro, to do away with these limitations. However, for the sake of simplicity, the travel along the linear axis would be assumed to be occurring at a uniform rate, synchronized with angular displacement. In other words, the pitch would be assumed constant, as in the case of the built-in helical interpolation. The radius would, of course, vary but the change in radius would be assumed to be at a uniform rate, with respect to angular displacement. Thus, the helix would be formed on the lateral surface of a right-angle cone.

A macro with this feature can be used for thread milling of taper pipe threads, with a modification to introduce pitch instead of number of loops. This is left as an exercise for the readers. However, it is possible to use the given macro even without any modification— one only has to calculate and then specify the number of turns (which can be in fraction also) by dividing the total depth (Z-travel) by pitch.

Consider a helix with a constant pitch and uniformly varying radius, having the following characteristics:

Number of loops (turns)	$= T$
Start radius	$= R_1$
End radius	$= R_2$
Total depth	$= d$
Plane of circular interpolation	$= XY$-plane
Linear axis	$= Z$-axis
XY-coordinates of helix axis	$= (X_0, Y_0)$
Coordinates of the start point	$= (X_1, Y_1, Z_1)$
Start angle, with positive X-axis	$= \theta_1$
Direction of toolpath	$= CCW$

The top view of such a helical toolpath is shown in Fig. 8.3, where four complete turns are shown, and the end radius is smaller than the start radius (though it does not affect the mathematical equation of the helix).

To have conformity with the way the built-in helical interpolation is used, the macro would be designed in such a manner that the start point of the helix would be at the current tool position, and the tool will remain at the end point of the helix in the end. In other words, the tool position at the time of calling the macro would automatically become the start point of the helix. This means that the center line (axis) of the

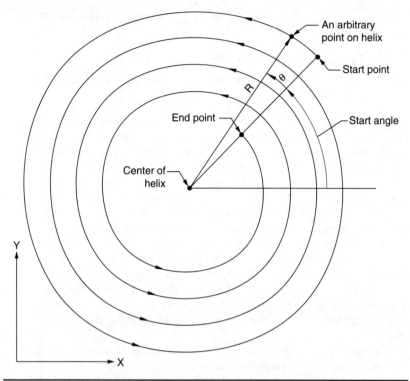

FIGURE 8.3   Top view of helical interpolation with varying radius.

helix automatically gets located if the start radius and the start angle are specified. Hence, its coordinates (X_0, Y_0) can be calculated.

The first task is to mathematically generate the equation of the helix. For this, the parametric equation of an arbitrary point on the helix would need to be written, with angle θ as a parameter:

$$X_0 = X_1 - R_1 \cos\theta_1$$

$$Y_0 = Y_1 - R_1 \sin\theta_1$$

$$R(\theta) = R_1 - \frac{R_1 - R_2}{360T}\theta$$

$$X(\theta) = X_0 + R\cos(\theta_1 + \theta)$$

$$Y(\theta) = Y_0 + R\sin(\theta_1 + \theta)$$

$$Z(\theta) = Z_1 + \frac{d}{360T}\theta$$

Now, the algorithm for this problem, suitable for WHILE_DO_END loop, can be written:

1. Determine the current tool position (X_1, Y_1, Z_1), through appropriate system variable.

2. Find out the XY-coordinates (X_0, Y_0) of the helix axis, using the given formulas.

3. Use the specified increment in angle, $\Delta\theta$, and assign $\theta = \Delta\theta$.

4. If θ is greater than 360T, go to step 9.

5. Calculate $R(\theta)$, $X(\theta)$, $Y(\theta)$, and $Z(\theta)$, using the given formulas.

6. Move to the calculated coordinates by G01, with the specified feedrate.

7. Increment angle $(\theta = \theta + \Delta\theta)$.

8. Go to step 4.

9. Return to the calling program.

Program number 8017 is based on this algorithm, which uses the local variables given in Table 8.2.

For helical interpolation with uniformly varying radius, this macro would need to be called in a manner such as

```
G65 P8017 T10 S50 E40 D30 A45 Q1 F60;
```

in the absolute mode, after placing the tool at the desired start point of the helix. Note that one need not necessarily be in G17 mode, as G01 remains unaffected by G17/G18/G19.

```
O8017 (VAR RAD HELICAL INTERPOLATION);
IF [#19 LT 0] THEN #3000 = 1 (NEGATIVE START RADIUS);
IF [#8 LT 0] THEN #3000 = 2 (NEGATIVE END RADIUS);
IF [#20 LT 0] THEN #3000 = 3 (NEGATIVE NUMBER OF TURNS);
```

> (Negative values for start radius, end radius, and number of turns are not allowed; these would

Parameters of the Helix	Selected Letter Addresses	Associated Variables
Number of turns	T	#20
Start radius	S	#19
End radius	E	#8
Total depth	D	#7
Start angle (with X-axis)	A	#1
Increment in angle	Q	#17
Feedrate	F	#9

TABLE 8.2 Parameters of Variable-Radius Helical Interpolation and the Selected Letter Addresses

generate respective alarms with the specified messages. Number of turns can have a fractional value also. For example, T = 1.1 refers to 396°)

```
#100 = #5041;
```
(Current X-position determined)

```
#101 = #5042;
```
(Current Y-position determined)

```
#102 = #5043;
```
(Current Z-position determined. **Step 1 complete**)

```
#103 = #100 - #19 * COS[#1];
```
(X-coordinate of helix axis calculated. Start angle can be negative also. It must be specified in degrees. Entire 360° range is permitted)

```
#104 = #101 - #19 * SIN[#1];
```
(Y-coordinate of helix axis calculated. **Step 2 complete**)

```
#105 = #17;
```
(Δθ set to the specified value, in degree, in the Q-word, which can be both positive and negative, for CCW and CW helices, respectively)

```
#106 = #105;
```
(θ set equal to Δθ. **Step 3 complete**)

```
WHILE [ABS[#106] LE [360 * #20]] DO 1;
```
(Exit from the loop if the magnitude of θ is greater than the total traversed angle. **Step 4 complete**)

```
#107 = #19 - [[#19 - #8] * ABS[#106] / [360 * #20]];
```
(R(θ) calculated)

```
#108 = #103 + #107 * COS[#1 + #106];
```
(X(θ) calculated)

```
#109 = #104 + #107 * SIN[#1 + #106];
```
(Y(θ) calculated)

```
#110 = #102 + [#7 * ABS[#106] / [360 * #20]];
```
(Z(θ) calculated. Both positive and negative values can be specified for the depth. A positive value moves the tool in the positive Z-direction, while going from the start point to the end point, whereas a negative value makes it move in the negative Z-direction. **Step 5 complete**)

```
G01 X#108 Y#109 Z#110 F#9;
```
(Move to the calculate coordinate with the specified feedrate. **Step 6 complete**)

```
#106 = #106 + #105;
```
(Angle θ incremented by the specified step value. The angle actually gets decremented if the step value is negative, which generates a clockwise helix. **Step 7 complete**)

```
END 1;
```
(End of WHILE loop. The tool stays at the end point of the helix. **Step 8 complete**)

M99;   (Return to the calling program.
                                        Step 9 complete)

It is left as an exercise for the readers to modify this macro to make it suitable for both absolute and incremental modes. In fact, it is pretty simple:

- In the beginning of the macro, read system variable #4003 to find out the mode (it stores 90 in absolute mode, and 91 in incremental mode).

- Replace G01 by G90 G01 in the macro, to force absolute mode.

- Command G91 at the end of the macro, if the first step revealed incremental mode.

The examples given above explain how a complex toolpath can be generated, if it is mathematically defined. There is no need to use a CAM software for such a requirement. Moreover, a CAM software may not provide as much flexibility as is possible with a macro. Some examples for a lathe are discussed next.

A complication in turning applications is that the depth of cut must be limited to its maximum permissible value, for a given cutting condition. Hence, it is not possible to obtain a complex turned shape just by moving the tool along the generated boundary. One has to make use of G71/G72/G73, where the part boundary is defined between P- and Q-block numbers. If G71/G72 type II cycles are not available on a particular machine, G73 would have to be used wherever the change in part diameter, along the axis of the part, is not monotonic. G71/G72 type I cycles require that there be monotonic increase or decrease in part diameter, along its axis.

Two cases are discussed below: one with monotonic increase, and the other with nonmonotonic increase. The monotonic case may be considered suitable for G71, whereas the nonmonotonic case uses G73. However, as explained in the next section, G71/G72 cannot be used for such applications.

8.4   Parabolic Turning

Reflectors are given parabolic shapes because a signal or light emitted at its focus becomes parallel to its axis after reflection, which makes transmission possible over a long distance. Similarly, if it is used for receiving signals from a long distance (which are nearly parallel rays), a signal receptor placed at its focus receives strong converging signals. Parabolic turning, however, is not a standard feature of a lathe, though it can be very easily done using a macro.

Consider the geometry shown in Fig. 8.4. It is assumed that the workpiece has an initial bore, and the girth of the internal turning tool is small enough to enter the hole, without any interference. If not, a central hole of suitable diameter would be required to be drilled first, before proceeding with parabolic turning. The method given here, however, cannot be directly used if the design does not permit existence of a central hole. In such a case, machining would have to be done using the *pattern repeating cycle* (G73) in facing operation (i.e., with *radial relief* = 0 and *axial relief* = as required), with a neutral facing tool. However, as a first approximation, a central hole can still be made, with a slotdrill, to have a flat surface at the bottom of the hole. The error introduced (the distance between the vertex of the parabola and the flat bottom) may be negligibly small. For example, it is 0.3175 mm for the dimensions given in Fig. 8.4.

The first task is to derive a mathematical equation for the parabola. It can be shown that the general equation in terms of the maximum diameter (M), bore diameter (B), and the depth (D) is

$$X^2 = \frac{M^2 - B^2}{D}\left(Z + \frac{M^2 D}{M^2 - B^2}\right)$$

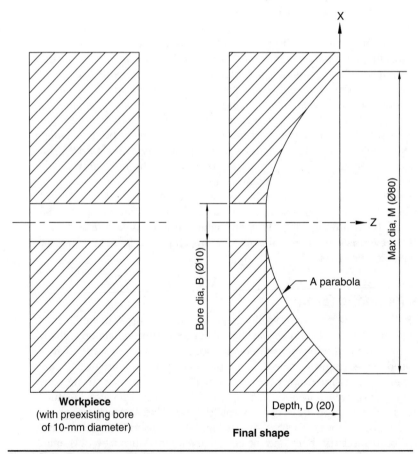

FIGURE 8.4   Turning a parabola.

One can verify that Z = 0 gives X = M, and Z = –D gives X = B. Note that this equation has been written in a form to suit *diameter programming* on a lathe. For the dimensions given in Fig. 8.4, this equation reduces to

$$X^2 = 315(Z + 20.3175)$$

However, instead of using this equation, the general format would be used in the macro.

Machining of this type, both internal and external, which involves bulk material removal, is usually done by G71, the *multiple turning cycle* (also called the *stock removal cycle*, or simply the *roughing cycle*) or G72, the *multiple facing cycle*, as these are the most efficient cycles. So, one may think of using, say, G71, for machining, where the given equations would be used for defining the parabolic curve between P- and Q-block numbers. (Note, however, that when the radial dimension of the material to be removed is more than its axial dimension, as in the present case where these values are 35 and 20 mm, respectively, G72 would be more efficient than G71.) An algorithm involving G71 can be written in the following manner:

1. Place the tool at the start point of G71.
2. Command G71 and place the tool at the start point of the parabola, that is, at X = M and Z = 0.
3. Use the specified axial step distance ΔZ, and assign Z = –ΔZ.
4. If |Z| is greater than D, which is $|Z_{max}|$, go to step 9.
5. Calculate X, using the given formula.
6. Move to the calculated coordinates by G01.
7. Change the Z-coordinate (Z = Z – ΔZ).
8. Go to step 4.
9. End of G71.
10. Finish by G70 with half the specified feedrate, and twice the current rpm, for a better finish.
11. Return to the calling program.

Unfortunately, the technique described above does not work. This is because though G71 and G72 do allow loops between the P- and Q-blocks, presence of a movement command in a loop is not allowed. Therefore, it is not possible to define the geometry of the part through a loop, for G71/G72. Moreover, even subprograms/macros cannot be called from within G71/G72 cycles (M98 is ignored, and G65/G66 alarm out). G73, the *pattern repeating cycle*, does not have these limitations, but it would be a highly inefficient cycle for such applications, as it would be cutting in the air most of the time.

Hence, the only way, in such a case, is to simulate the roughing operation of G71 and G72 by calling G90 and G94, respectively, in a loop. The step-removal operation of G71/G72 can be simulated by moving the tool along the desired profile, defined as tiny straight line segments. An algorithm for simulating G72 (which is more efficient compared to G71, for the given dimensions) is given below:

Roughing Operation

1. Place the tool at Z = 0, with 1 mm clearance from the hole, i.e., at X = B – 2.
2. Assign Z = – <specified roughing depth of cut>.
3. If |Z| is greater than D, go to step 9.
4. Calculate X, using the given formula.
5. Command G94, with the calculated coordinates and the specified feedrate as its arguments.
6. Shift the tool axially, to the Z-level of the machined surface, which would be the start point of the next G94 cycle.
7. Change the Z-coordinate (Z = Z – <specified roughing depth of cut>).
8. Go to step 3.

Step-Removal Operation

1. Place the tool at the inner edge of the parabola, that is, at (B, –D).
2. Assign Z = –D.
3. If Z is greater than 0, go to step 16.
4. Calculate X, using the given formula.
5. Move to the calculated coordinates by G01.
6. Change the Z-coordinate (Z = Z + <specified step distance in Z>).
7. Go to step 11.
8. Return to the calling program.

Note that if the depth of the parabola is not an exact multiple of the specified step distance in Z, the tool will not exactly reach the outer edge of the parabola. This would leave a small kink at the edge. This can be removed by moving the tool by some distance along the tangential direction at Z = 0. It can be shown that the increments in X and Z, at Z = 0, are related by the following formula:

$$\Delta X = \frac{M^2 - B^2}{2MD} \Delta Z$$

Parameters of the Parabola	Selected Letter Addresses	Associated Variables
Maximum diameter	M	#13
Bore diameter	B	#2
Depth of parabola	D	#7
Step distance in Z	Q	#17
Roughing depth of cut	R	#18
Feedrate	F	#9

TABLE 8.3 Parameters of Parabolic Turning and the Selected Letter Addresses

Thus, an additional tangential move, using the given formula, is desirable before returning to the calling program.

Program number 8018 is based on this algorithm, using the local variables given in Table 8.3. The toolpath for this program is shown in Fig. 8.5. For the geometry given in Fig. 8.4, this program would need to be called in a manner such as

```
G00 X0 Z2;
G65 P8018 M80 B10 D20 Q0.1 R0.5 F60;
```

The initial positioning of the tool at (X0, Z2) avoids a possible interference between the tool and the workpiece or the machine body, when the tool makes the first move inside the macro. Roughing is done at the current rpm and the specified feedrate, whereas finishing is done at twice the current rpm and half the specified feedrate. This gives a better surface finish.

`O8018 (PARABOLIC TURNING WITH G94);`	
`#105 = #5041;`	(Stores the initial X-position of the tool)
`#106 = #5042;`	(Stores the initial Z-position of the tool)
`G00 X[#2 - 2] Z0;`	(Tool placed at the start point of the first G94 cycle, leaving a clearance of 1 mm in the radial direction. **Step 1 complete**)
`#100 = ABS[#17];`	(ABS function would accept both positive and negative values for step distance in Z)
`#101 = -#18;`	(The first step change in Z. **Step 2 complete**)
`WHILE [ABS[#101] LE ABS[#7]] DO 1;`	(Exit from the loop if the entire depth is roughed out. **Step 3 complete**)

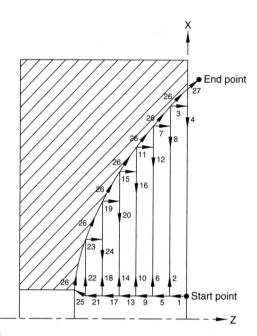

Note:
1. The sequence of toolpath is numbered.
2. Rapid motions: 1, 4, 5, 8, 9, 12, 13, 16, 17, 20, 21, 24.
3. Feed motions: 2, 3, 6, 7, 10, 11, 14, 15, 18, 19, 22, 23, 25, 26, 27.
4. Toolpath 1 to 24, caused by executing G94 in a loop (along with a positioning move at the end of each G94 cycle), simulates the roughing operation of G72 (without finishing allowances), with the difference that the retractions (3, 7, etc.) are not at 45°.
5. Toolpath 25 is a positioning move, after the completion of all G94 cycles.
6. Toolpath 26 consists of very small straight-line path segments along the defined parabolic profile. It removes the steps that are nearly triangular in shape, created along the parabolic profile, by toolpath 1 to 24. Thus, it simulates the step-removal operation of G72. The accuracy of machining would, obviously, depend on the chosen value of Z-increment for generating toolpath 26.
7. Toolpath 27 is a straight-line move along the tangential direction, for removing a possible kink at the outer edge of the parabola.
8. The final rapid move, to the position of the tool immediately before calling the macro, is not shown.

Figure 8.5 Toolpath for turning a parabola.

```
#102 = [[#13 * #13 * ABS[#7]] / [#13 * #13 - #2 * #2]];
                          (Intermediate calculation)
#102 = #101 + #102;       (Intermediate calculation)
#102 = [[#13 * #13 - #2 * #2] / ABS[#7]] * #102;
                          (Intermediate calculation)
#102 = SQRT[#102];        (X calculated. Step 4 complete)
G94 X#102 Z#101 F#9;      (Roughing with G94. Step 5 complete)
```

```
G00 Z#101;
```
(Tool positioning for the next G94 cycle. **Step 6 complete**)

```
#101 = #101 - #18;
```
(Z decremented by the specified roughing depth of cut. **Step 7 complete**)

```
END 1;
```
(End of loop. **Step 8 complete**)

```
#103 = #4119;
```
(Current rpm stored)

```
G01 X#2 Z[-ABS[#7]] F[#9 / 2] S[#103 * 2];
```
(Tool placed at the inner edge of the parabola. **Step 9 complete**)

```
#101 = -ABS[#7];
```
(Z-coordinate of the inner edge of the parabola stored. **Step 10 complete**)

```
WHILE [#101 LE 0] DO 1;
```
(Exit from the loop if the outer edge of the parabola is reached. **Step 11 complete**)

```
#102 = [[#13 * #13 * ABS[#7]] / [#13 * #13 - #2 * #2]];
```
(Intermediate calculation)

```
#102 = #101 + #102;
```
(Intermediate calculation)

```
#102 = [[#13 * #13 - #2 * #2] / ABS[#7]] * #102;
```
(Intermediate calculation)

```
#102 = SQRT[#102];
```
(X calculated. **Step 12 complete**)

```
X#102 Z#101;
```
(Linear interpolation to the calculated coordinates. **Step 13 complete**)

```
#101 = #101 + #100;
```
(Z incremented by the specified step distance. **Step 14 complete**)

```
END 1;
```
(End of loop. **Step 15 complete**)

```
#104 = [[#13 * #13 - #2 * #2] / [2 * #13 * ABS[#7]] * 2];
```
(Calculating ΔX, corresponding to $\Delta Z = 2$ mm, at Z = 0)

```
U#104 W2;
```
(Additional tangential move at the outer edge of the parabola for removing any possible kink)

```
G00 X#105 Z#106;
```
(Rapid move to the original tool position)

```
S#103 F#9;
```
(Original rpm and feedrate restored)

```
M99;
```
(Return to the calling program. **Step 16 complete**)

Though the macro O8018 does not have the feature of finishing allowance of G72, it can be easily incorporated into it. One only has to subtract the desired X-finishing allowance from the calculated X-values, and add Z-finishing allowance to the calculated Z-values, in roughing as well as in step-removal operations. Finally, the step-removal operation would be repeated without adding/subtracting finishing allowances, with the same tool or a different tool, if desired. It is also possible to introduce two more local variables for passing the values for X- and Z-finishing allowances to the macro, through the macro-calling block. It is left as an exercise for the readers to develop such a macro.

8.5   Turning a Sine Curve

Consider the part shown in Fig. 8.6, which is to be machined from a cylindrical workpiece. The lateral surface of this part is defined by a sine curve. An additional complexity in this case is that the diameter of the part does not have monotonic increase or decrease along the length. The diameter first increases, then decreases and again increases, in the negative Z-direction. This makes the roughing operation by G90/G94 difficult to incorporate, which was easily done in the previous example of parabolic turning. Hence, the only way is to make use of G73 cycle. In fact, one simply has to define the geometry of the profile between P- and Q-block numbers of G73.

As discussed earlier, this technique cannot be used with G71 and G72, because a loop cannot have movement G-codes for defining the profile for G71/G72. G73, however, does not have this limitation. The only problem, in fact a serious one, with G73 is that it is very inefficient for this type of job, where most of the feed motion would be in air, wasting a lot of time. However, in the absence of a suitable method, one has to use G73 that will at least produce the correct part. Ideally, G73 should be used where the workpiece already has a preformed shape, requiring nearly equal amount of material removal over its entire length, to get the exact shape. In other words, it is mainly suitable for machining of cast or forged parts, for which no other machining cycle would be as efficient. Every canned cycle has been designed with some specific purpose.

We are using the terms *cycle* and *canned cycle* interchangeably. Broadly speaking, if a G-code brings back the tool to its initial position after machining, it is called a cycle. If a cycle also involves some calculations, it is usually referred to as a canned cycle. Thus, G90 is a cycle whereas G71 is a canned cycle. Fanuc manuals, however, refer

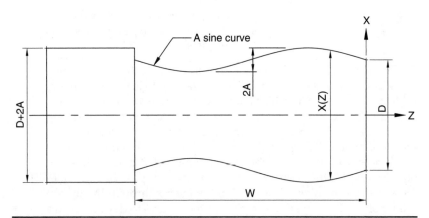

Figure 8.6   Turning a sine curve.

to all cycles as canned cycles. For this reason we are not differentiating between the two terms in this text.

Coming back to the original problem, an algorithm for turning a sine curve is given below:

1. Place the tool at the start point of G73.

2. Command G73 and place the tool at the start point of the sine curve, that is, at X = D and Z = 0.

3. Use the specified axial-step distance ΔZ, and assign Z = $-\Delta Z$.

4. If $|Z|$ is greater than W, which is $|Z_{max}|$, go to step 9.

5. Calculate X, using the equation of the sine curve:

$$X = D + A \sin\left(\frac{360|Z|}{W}\right)$$

6. Move to the calculated coordinates by G01.

7. Change the Z-coordinate (Z = Z − ΔZ).

8. Go to step 4.

9. End of profile definition of G73.

10. Finish by G70 with half the specified feedrate, and twice the current rpm, for a better finish.

11. Return to the calling program.

Program number 8019, which uses the local variables given in Table 8.4, is based on this algorithm.

Axial relief has to be kept zero in the case being considered, as the geometry of the part requires that there be no axial shift in the subsequent cutting passes of G73 (because axial shift would result in interference between the toolpath and the part geometry. In fact, for the same reason, even the Z-finishing allowance for the G73 cycle has to

Parameters of the Sine Curve	Selected Letter Addresses	Associated Variables
Average diameter	D	#7
Wavelength of sine curve	W	#23
Amplitude of sine curve	A	#1
Step distance in Z	Q	#17
Roughing depth of cut	R	#18
Feedrate	F	#9

TABLE 8.4 Parameters of Sine Curve Turning and the Selected Letter Addresses

be kept zero in this case). In this example, the cutting passes are required to shift only in the radial direction, to gradually approach the defined profile.

The radial relief in the present case can be chosen to be equal to the maximum depth of the "valley" (i.e., equal to 2A). However, with this radial relief, the first pass of G73 would only touch the workpiece at a location where the part has minimum diameter, without removing any material. Since it is desirable that material be removed even in the first cutting pass, radial relief should be taken equal to (2A − <desired depth of cut>).

Though the macro O8019 allows specification of roughing depth of cut, G73 does not use this information directly. Instead, one has to specify the number of cutting passes (R-word in the first block of G73), for the chosen radial/axial relief (U-word/W-word in the first block of G73), such that the resulting depth of cut is as desired.

Recall that G73 offsets the first cutting pass, with respect to the defined profile (shifted by finishing allowances, if specified, along the corresponding directions), by the specified radial- and axial-relief amounts, in radial and axial directions, respectively. The last cutting pass exactly follows the defined profile (plus finishing allowances). Since the total number of passes is the value specified in the R-word, there would be (<R-value> − 2) intermediate passes. All the passes are automatically evenly spaced, and approach the last pass (i.e., the defined profile plus finishing allowances) gradually. Therefore, the spacing between two consecutive passes determines the depth of cut in a pass.

The required number of passes (R-value), to ensure that the maximum depth of cut in each pass is equal to or smaller than the specified depth of cut, would be equal to

$$\frac{radial\,relief}{specified\,depth\,of\,cut} + 1$$

Since, in general, this expression would evaluate to a real number, the next (higher) whole number should be specified as the R-value. Note that the actual depth of cut in the example being considered would not remain constant. It would vary between zero (where no cutting is involved) and the specified depth of cut (where the part has minimum diameter).

```
O8019 (TURNING A SINE CURVE);
#105 = #5041;                       (Stores the initial X-position of the tool)
#106 = #5042;                       (Stores the initial Z-position of the tool)
G00 X[#7 + 2 * #1 + 4] Z2;          (Tool placed at the start point of G73
                                     cycle, leaving a clearance of 2 mm in
                                     radial as well as axial direction, with
                                     respect to the workpiece. Step 1
                                     complete)
```

```
#100 = ABS[#17];
```
(ABS function would accept both positive and negative values for step distance in Z)

```
G73 U[2 * #1 - #18] W0 R[FUP[[[2 * #1 - #18] / #18] + 1]];
G73 P1 Q2 U0.1 W0 F#9;
```
(X-finishing allowance chosen to be 0.1 mm)

```
N1 G00 X[#7 + 4] Z0;
G01 X#7 Z0;
```
(Tool placed at the start point of the sine curve. This motion has been split into two moves because the tool must approach the workpiece at feedrate, and a single feed motion would waste a lot of time. Modify the previous command if the depth of cut is more than 2 mm. **Step 2 complete**)

```
#101 = -#100;
```
(The first step change in Z. **Step 3 complete**)

```
WHILE [ABS[#101] LE ABS[#23]] DO 1;
```
(Exit from the loop if the sine curve gets defined completely, i.e., up to Z = – W. **Step 4 complete**)

```
#102 = [#7 + #1 * SIN[360 * ABS[#101] / ABS[#23]]];
```
(X calculated. **Step 5 complete**)

```
X#102 Z#101;
```
(Linear interpolation to the calculated coordinates. **Step 6 complete**)

```
#101 = #101 - #100;
```
(Z decremented by the specified step distance. **Step 7 complete**)

```
END 1;
```
(End of loop. **Step 8 complete**)

```
N2 X[#7 + 2 * #1];
```
(End of profile definition. The last segment, a radial movement up to the initial workpiece diameter, is generally given to ensure a better machined surface at the end. However, in the present case, where possibly a neutral tool would be used, this step may be considered unnecessary, and one can simply write N2 END 1. **Step 9 complete**)

```
#103 = #4119;
```
(Current rpm stored)

```
G70 P1 Q2 F[#9 / 2] S[#103 * 2];
```
(Finishing by G70. Make the necessary change here if a different tool is desired to be used for finishing. **Step 10 complete**)

```
G00 X#105 Z#106;
```
(Tool sent back to its original position)

```
F#9 S#103;
```
(The original feedrate and rpm restored)

```
M99;
```
(Return to the calling program. **Step 11 complete**)

This macro, to obtain the shape of Fig. 8.6, can be called in a manner such as

```
. . .
G98 G00 X100 Z0;
G65 P8019 D30 W65 A3 Q0.1 R0.5 F60;
. . .
```

The shank of a right-hand turning tool is likely to interfere with the geometry of this part, unless the tip angle is very small (say, 35°), and the amplitude to the length ratio of the sine curve is also small. This necessitates the use of a neutral tool. If a diamond-shaped insert is used in a neutral shank, the square edge at the left end of the sine curve cannot be obtained by this method. (There would be overcutting at half the tip angle on the left side.) One cannot use a round insert, because it has to be used with radius compensation (since the point on the tool that cuts the material continuously changes its position), and radius compensation cannot be used inside G71–G73 cycles (which is a limitation of these cycles).

CHAPTER 9

Parametric Programming

9.1 Introduction

The term "parametric programming" is not very clearly defined. Different people use it to mean different things. For example, some people use it in the sense we are using the term "macro programming." They would say that Fanuc Custom Macro B, Fadal Macro, Okuma User Task, etc. are the available programming languages for parametric programming. On the other hand, some people use this term in its mathematical sense, that is, for a program written in terms of parameters, which can be made to do different things just by changing the values of the parameters of the program. In other words, the technique of writing a single program for a *family of parts* is referred to as parametric programming. This definition has a wider acceptance and hence, adopted in this text.

Drilling of bolt holes on flanges, having different number of holes, different depth and angular orientation of holes, different pitch circle diameter, etc. is a good example of an application of parametric programming technique. All the flanges may have different dimensions, but similar characteristics. Thus, these belong to the same family of parts. In such cases, instead of writing different programs for different flanges, it is much better to write a single program in terms of (variable) parameters, and call it with the desired values for the parameters. Such a program would necessarily be a macro, though not all macros can be called parametric programs. For example, a macro for a user-designed peck-drilling cycle for very deep holes, which has gradually reducing peck lengths in subsequent pecks, does not qualify to be called a parametric program. A parametric program must relate to a family of parts. Thus, parametric programming is a subset of macro programming.

Incidentally, the term "parameter" used here is not at all related to machine parameters (which are also referred to as system parameters, control parameters, or simply parameters) that decide the default settings of a machine. (If a change in some default setting is needed, the corresponding machine parameters would need to be altered. This is usually done manually, in MDI mode. However, it is also possible to change these automatically, through any program, including a parametric program. This issue is discussed in Chap. 13.) The remaining part of this chapter is devoted to examples of parametric programs for both lathe and milling machines.

9.2   Locator Pin

Consider the locator pin shown in Fig. 9.1. Out of the seven dimensions that define the geometry of the pin, only two dimensions are fixed. The remaining five dimensions are given in terms of variables A, B, C, D, and R. The values of these variables are given in Table 9.1, for five pins of different dimensions. There is no need to write five

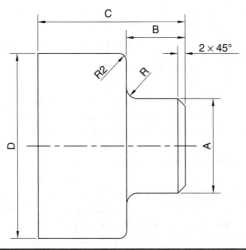

Figure 9.1   Locator pin.

Pin Number	A	B	C	D	R
1	10	10	15	25	2
2	15	10	15	30	2
3	20	15	25	35	3
4	25	15	25	40	3
5	30	20	35	45	4

Table 9.1   Dimensions of the Locator Pin Shown in Fig. 9.1

different programs for the five pins. A single parametric program O8020 would produce all five pin types (as well as pins with any other dimensions, as long as the overall geometry is the same), when called with the desired values for parameters A, B, C, D, and R that correspond to local variables #1, #2, #3, #7, and #18, respectively, in program O8020. Thus, G65 P8020 A10 B10 C15 D25 R2 would produce pin number 1, G65 P8020 A15 B10 C15 D30 R2 would produce pin number 2, and so on.

The following assumptions have been made while writing the macro:

1. The calling program is in millimeter mode (G21).

2. The initial diameter of the workpiece is 50 mm, for all the pins.

3. The workpiece zero point is at the center of the right face of the workpiece. (In fact, all the lathe programs in this text assume the same location for the workpiece zero point.)

4. The workpiece requires a facing of 0.5 mm, in one pass. (A tailstock is not being used to support the workpiece.) Note that this would result in the right face of the part being at $Z = -0.5$ mm, since $Z = 0$ is assumed to lie on the right face of the workpiece.

5. The feedrate and the spindle speed (or CSS, the constant surface speed in G96 mode) in the calling program are correct and would be used for roughing. For finishing, feedrate would be halved and spindle speed/CSS would be doubled.

6. Tool number 1 with offset number 1 would be used for roughing, and tool number 3 with offset number 3 would be used for finishing.

7. G71 (Type 1) and G70 would be used for roughing and finishing, respectively.

8. Part-off operation is not needed (would be done separately, using, say, G75), for which 3 mm extra length should be provided on the larger diameter.

These assumptions are not really a limitation of the given macro. One always has to first decide what the macro is supposed to do, which is what has been done here. If some of these assumptions are not quite suitable in a certain case, the macro would need to be modified accordingly. In general, modifying in a macro is much simpler than writing a new macro. In fact, one should always start with a simple and easily achievable goal. The complexities should be incorporated one by one, only after the basic macro is ready and tested. Trying to develop a complex macro in the very first attempt is a source of confusion, leading to programming errors. Such a programming style was recommended earlier in this text. It is being reiterated here because it is so important.

This is not an example of a macro involving a complex logic. Therefore, a flowchart or an algorithm is not really necessary. It is, however, recommended to at least write down how the toolpath is to be generated, which would streamline the thought process:

1. Send the turret to its home position, and select tool number 1 with offset number 1.

2. Place the tool at (X54 Z2), and face up to Z–0.5 using G94, in a single pass.

3. Place the tool at (X50 Z2), and rough out using G71, leaving suitable finishing allowances.

4. Send the turret to its home position, and select tool number 3 with offset number 3.

5. Place the tool at (X50 Z2), and finish by G70, at reduced feedrate and increased rpm/CSS, with radius compensation. (Since it is a right-hand turning tool, the *tool-tip number* on a rear-type lathe would be 3 that has to be entered in the third row of the geometry **or** wear offset table. The radius of the tool tip is entered in the geometry offset table.)

The defined profile, which is same for both G71 and G70, is shown in Fig. 9.2.

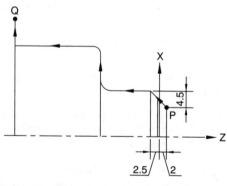

Note:
1. Refer to Fig. 9.1 for dimensions not shown.
2. The profile definition starts at P and ends at Q.
3. The X- and Z-coordinates of P are (A – 9) and 2, respectively.
4. The X- and Z-coordinates of Q are 54 and –(C + 3.5), respectively.
5. The extended toolpath at the start of the profile is at the angle of the chamfer (45°). This makes a better-quality corner.
6. The Z-coordinate of Q includes 3 mm extra length on the larger diameter, which is needed for part-off operation.

FIGURE 9.2   Defined profile for G71 and G70, for the pin of Fig. 9.1.

```
O8020 (LOCATOR PIN);
G28 U0;
G28 W0;




T0101;



G00 Z2;
X54;




M03;

M08;

G94 X-4 Z-0.5;







G00 X50 Z2;


G71 U2 R0.5;


G71 P10 Q20 U0.2 W0.1;







N10 G00 X[#1 - 9];





G01 X#1 Z-2.5;
Z-[#2 + 0.5 - #18];
G02 X[#1 + 2 * #18] Z-[#2 + 0.5] R#18;

G01 X[#7 - 4];

G03 X#7 Z-[#2 + 2.5] R2;
```

(X-home)

(Z-home. Separate X- and Z-homing eliminates the possibility of any interference between the moving turret and the machine body, such as the tailstock)

(Tool number 1, with offset number 1, brought in the cutting position. **Step 1 complete**)

(Positioning move)

(Tool placed at the start point of G94. Here also, two separate motions are given to eliminate the possibility of interference)

(Clockwise rotation of the spindle starts, at the currently active rpm)

(Coolant starts. Usually, only one M-code in a line is permitted)

(Facing up to below the axis of the workpiece removes any kink at the center. In the absence of an F-word, the active feedrate, at the time of calling the macro, would be used for facing. **Step 2 complete**)

(Tool placed at the start point of G71 cycle. Z2 is redundant in this command)

(2 mm depth of cut, and 0.5 mm radial retraction chosen. Modify these values, if not considered suitable)

(The profile definition starts at N10 block and ends at N20 block. The X- and Z-finishing allowances are 0.2 and 0.1 mm, respectively. Modify these values, if not considered suitable. In the absence of an F-word, the previous feedrate would be used for roughing)

(Profile definition starts here. G71 (Type 1) does not allow a Z-word in this block. It automatically uses the current Z-position, for locating the first point on the profile, point P in Fig. 9.2)

(Making chamfer)
(Straight turning on smaller diameter)

(Making concave fillet)

(Straight facing toward the larger diameter)

(Making convex fillet)

`G01 Z-[#3 + 3.5];`	(Straight turning on larger diameter, providing 3 mm extra length for part-off operation)
`N20 X54;`	(Straight facing to clear off the workpiece. Profile definition ends here, with 2 mm radial clearance from the workpiece, which is point Q in Fig. 9.2. **Step 3 complete**)
`G28 U0;`	(X-home)
`G28 W0;`	(Z-home)
`T0303;`	(Tool number 3 with offset number 3 selected for finishing. **Step 4 complete**)
`G42 G00 Z2;`	(Radius compensation invoked. It was not used earlier because radius compensation is ignored by the canned cycles, G71–G76)
`X50;`	(Some controls require that the first two moves, after invoking radius compensation, be along the two axes, separately, before reaching the start point of the profile, otherwise there might be some error in tool positioning at the start point. On Fanuc control, however, a single start-up move, such as G42 G00 X50 Z2, is sufficient. Here, the main reason for splitting it into two moves is interference avoidance)
`G70 P10 Q20 F[#4109 / 2] S[#4119 * 2];`	
	(Finishing by half the current feedrate and twice the current rpm/CSS. Modify if desired. **Step 5 complete**)
`G40 U4 W2;`	(Radius compensation canceled. As a rule of thumb, cancelation must be accompanied by adequate outward movement, otherwise, the tool may move toward the workpiece during cancellation. Even if compensation is not explicitly canceled by G40, it would be canceled automatically by M30 which resets the control)
`M05;`	(Spindle stops)
`M09;`	(Coolant stops)
`G28 U0;`	(X-home)
`G28 W0;`	(Z-home)
`F[#4109 * 2] S[#4119 / 2];`	(The original feedrate and rpm/CSS restored)
`M99;`	(Return to the calling program)

For making pin number 1, this macro may be called by a program such as the one given below:

```
G21 G97 G98;                    (Change to G96 for CSS, and G99 for
                                feedrate in millimeters per revolution)
F60 S1000;                      (Specify roughing feedrate and spindle
                                speed)
G65 P8020 A10 B10 C15 D25 R2;
M30;
```

The given macro can make pins with any dimension, defined in terms of parameters A, B, C, D, and R. However, if the requirement is to make pins of only the five given types, and the operator is not expected to remember the meanings of the associated letter addresses and/or the exact dimensions of the pins, one can write a modified main program that would produce the desired type of pin just by entering the appropriate pin number in the beginning of the program. Such a program can be called a program for a family of parts, in the true sense:

```
#100 = 1;                       (Specify pin number here. One can also
                                use a local variable such as #1 here. This
                                would be different from variable #1
                                used inside the macro, as the local vari-
                                ables of the main program and those of
                                the macro called by it belong to level 0
                                and level 1, respectively)
IF [[#100 LT 1] OR [#100 GT 5]] THEN #3000 = 1 (ILLEGAL
PIN NUMBER);
                                (Any value other than 1, 2, 3, 4, and 5
                                would alarm out, terminating the pro-
                                gram execution)
G21 G97 G98;
F60 S1000;
IF [#100 EQ 1] GOTO 1;          (Branches out to N1 for producing pin 1)
IF [#100 EQ 2] GOTO 2;          (Branches out to N2 for producing pin 2)
IF [#100 EQ 3] GOTO 3;          (Branches out to N3 for producing pin 3)
IF [#100 EQ 4] GOTO 4;          (Branches out to N4 for producing pin 4)
IF [#100 EQ 5] GOTO 5;          (Branches out to N5 for producing pin 5)
N1 G65 P8020 A10 B10 C15 D25 R2;
                                (Macro call with pin-1 parameters)
GOTO 6;
N2 G65 P8020 A15 B10 C15 D30 R2;
                                (Macro call with pin-2 parameters)
GOTO 6;
N3 G65 P8020 A20 B15 C25 D35 R3;
                                (Macro call with pin-3 parameters)
GOTO 6;
N4 G65 P8020 A25 B15 C25 D40 R3;
                                (Macro call with pin-4 parameters)
GOTO 6;
```

```
N5 G65 P8020 A30 B20 C35 D45 R4;
```
                     (Macro call with pin-5 parameters)

```
N6 M30;
```

9.3   Bolt Holes on a Flange

Consider the flange shown in Fig. 9.3. It can be described in terms of the following parameters. (The corresponding local variables, as per argument specification 1, are indicated in brackets.)

X (#24) = X-coordinate of the center of the pitch circle
Y (#25) = Y-coordinate of the center of the pitch circle
Z (#26) = depth of holes
D (#7) = pitch circle diameter
H (#11) = number of holes
S (#19) = start angle (angle of the first hole, with the X-axis)
R (#18) = R-point level for the drilling cycle
F (#9) = feedrate

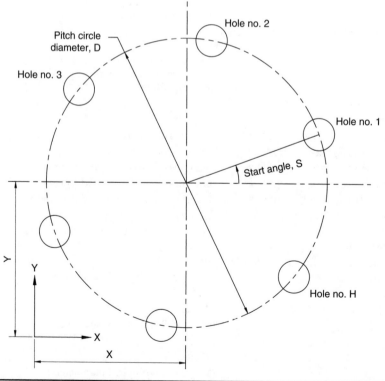

Figure 9.3   Bolt holes on a flange.

To keep things simple, G81 would be used for drilling holes, at the current spindle speed. The readers are, however, encouraged to introduce additional local variables for these, which is being left as an exercise. For a drilling cycle other than G81, information regarding dwell value (for G82) or peck distance (for G73/G83) would also be needed, necessitating the use of two more local variables for assigning values for these. Note that there is no harm in specifying a dwell value or a peck distance for a canned cycle that does not use these features. It would simply ignore such arguments. Hence, the canned-cycle block can have the same argument list, irrespective of whether it refers to G81, G82, G73, or G83.

Though a macro can be written for machining a flange with any values for the chosen parameters, a family of flanges, as described in Table 9.2, would be considered here. Note that X and Y are not really the parameters of a flange. Their values depend on how the flange has been clamped on the machine table, with respect to the active work coordinate system (G54, G55, etc.). Hence, the same values for these have been taken for all the flanges, as these values have no significance with regard to the type of a flange. It was necessary to include these variables for the sake of generality. Similarly, the start angle may not have any relevance. Hence, 0° has been taken in all cases. It has been introduced to take care of some special requirement.

We would assume the following (though it is not difficult to relax these assumptions, it has been left as an exercise for the readers):

1. The calling program uses absolute coordinates (G90).
2. The proper tool is held by the spindle that rotates with proper rpm, before calling the macro.
3. Tool length compensation has been incorporated in the calling program.

Flange Number	X	Y	Z	D	H	S	R	F
1	150	100	5	50	4	0	1	60
2	150	100	10	100	6	0	1	50
3	150	100	15	150	8	0	1.5	40
4	150	100	20	200	10	0	2	30
5	150	100	20	300	12	0	2	30

TABLE **9.2** Dimensions of the Flange Shown in Fig. 9.3

The following algorithm can be used for developing a macro (O8021) for this problem:

1. Set arguments for G81, with K0 (with K0, the information gets stored without drilling a hole).

2. Calculate the angle between consecutive holes, by dividing 360 by the number of holes.

3. Set <hole counter> = 1.

4. If the hole counter is greater than the number of holes, go to step 8.

5. Specify XY-coordinates of the first/next hole. (This would make the tool move to the specified coordinates where a hole would be drilled by G81.)

6. Increment hole counter by 1.

7. Go to step 4.

8. Cancel canned cycle.

9. Return to the calling program.

```
O8021 (BOLT HOLES ON A FLANGE);
G81 Z[ABS[#26]] R#18 F#9 K0;
```
	(Arguments of G81 specified. **Step 1 complete**)
`#100 = [360 / #11];`	(Angle between subsequent holes calculated. A negative value for the number of holes, H, is also permitted that would drill the holes in the clockwise direction. **Step 2 complete**)
`#101 = 1;`	(Hole counter initialized. Its value indicates the hole number that is to be drilled next. 1 and #11 correspond to the first and the last holes, respectively. **Step 3 complete**)
`WHILE [#101 LE ABS[#11]] DO 1;`	(Jump out of the loop if the hole counter exceeds the specified number of holes, i.e., when all the holes have been drilled. **Step 4 complete**)
`#102 = #19 + #100 *[#101 - 1];`	(Calculates the angular position of the first/next hole with respect to the X-axis, i.e., the CCW angle subtended at the center of the pitch circle by the position where the next drilling is to be done. A negative value for the start angle, #19, is also permitted)
`#103 = #24 + [#7 / 2]COS[#102];`	(X-coordinate of the first/next hole calculated)

```
#104 = #25 + [#7 / 2] SIN[#102];
```
(Y-coordinate of the first/next hole calculated)

```
X#103 Y#104;
```
(First/next hole drilled by G81 that is currently active. As long as it is not canceled by, say, G80, every axis displacement would cause a hole to be drilled at the new position, using G81. **Step 5 complete**)

```
#101 = #101 + 1;
```
(Hole counter incremented by 1, for drilling the next hole. **Step 6 complete**)

```
END 1;
```
(End of the current loop. Jumps to the WHILE block, to check the condition for executing the next loop. **Step 7 complete**)

```
G80;
```
(G81 canceled. A canned cycle gets canceled also by Group 1 G-codes such as G00 and G01. **Step 8 complete**)

```
M99;
```
(Return to the calling program. **Step 9 complete**)

If some error traps are desired to be added to the macro, insert the following blocks in the beginning of the program:

```
IF [#24 EQ #0] THEN #3000 = 1 (PCD CENTER X NOT SPECIFIED);
IF [#25 EQ #0] THEN #3000 = 2 (PCD CENTER Y NOT SPECIFIED);
IF [#26 EQ #0] THEN #3000 = 3 (HOLE DEPTH NOT SPECIFIED);
IF [#7 EQ #0] THEN #3000 = 4 (PCD NOT SPECIFIED);
IF [#11 EQ #0] THEN #3000 = 5 (NO. OF HOLES NOT SPECIFIED);
IF [#19 EQ #0] THEN #3000 = 6 (START ANGLE NOT SPECIFIED);
IF [#18 EQ #0] THEN #3000 = 7 (R-POINT NOT SPECIFIED);
IF [#9 EQ #0] THEN #3000 = 8 (FEEDRATE NOT SPECIFIED);
```

For making flange number 1, this macro may be called by a program such as the one given below:

```
G21 G94;
```
(Change to G20 for inch mode, and G95 for feedrate in feed per revolution)

```
G91 G28 Z0;
G28 X0 Y0;
```
(Tool magazine sent to the home position, for changing tool)

```
M06 T01;
```
(Desired tool placed in the spindle)

```
G90;
```
(Absolute coordinate mode selected)

```
G00 X0 Y0;
```
(XY positioning. Specify different coordinate values, if needed, e.g., when the center of the pitch circle is too far from the XY-datum)

```
G43 H01 Z50;
```
(Tool length compensation invoked. Tool placed at Z = 50 mm. It defines the initial Z-level for G81)

```
M03 S1000;
```
(Specify appropriate CW spindle speed)

```
M08;
```
(Coolant starts)

```
G65  P8021  X150  Y100  Z5  D50  H4  S0  R1  F60;
M09;                      (Coolant stops)
M30;
```

The next task is to modify the calling program so that the desired type of flange is produced just by specifying its number in the beginning of the program. This would obviate the need for referring to the drawing/data table for obtaining the dimensions of different flanges. Moreover, there would be no need for understanding the significance of each letter address in the G65 block. Such a program, with an additional provision to specify values for the coordinates of the center of the pitch circle and the start angle (since the values given in Table 9.2 may not apply in all cases), is given below:

`#100 = 1;`	(Specify flange number)
`#101 = 150;`	(Specify X-coordinate of the center of the pitch circle. One way to determine the value is to manually move the spindle to the desired center of the pitch circle, and then read the coordinate display)
`#102 = 100;`	(Specify Y-coordinate of the center of the pitch circle)
`#103 = 0;`	(Specify start angle)
`IF [[#100 LT 1] OR [#100 GT 5]] THEN #3000 = 1 (ILLEGAL FLANGE NUMBER);`	
	(Any value other than 1, 2, 3, 4, and 5 would alarm out, terminating the program execution)
`G21 G94;`	
`G91 G28 Z0;`	
`G28 X0 Y0;`	
`M06 T01;`	(Modify T01 if a tool number other than 1 is to be used)
`G90;`	
`G00 X#101 Y#102;`	(XY-positioning at the center of the pitch circle)
`G43 H01 Z50;`	
`M03 S1000;`	
`M08;`	
`IF [#100 EQ 1] GOTO 1;`	(Branches out to N1 for producing flange 1)
`IF [#100 EQ 2] GOTO 2;`	(Branches out to N2 for producing flange 2)
`IF [#100 EQ 3] GOTO 3;`	(Branches out to N3 for producing flange 3)
`IF [#100 EQ 4] GOTO 4;`	(Branches out to N4 for producing flange 4)
`IF [#100 EQ 5] GOTO 5;`	(Branches out to N5 for producing flange 5)
`N1 G65 P8021 X#101 Y#102 Z5 D50 H4 S#103 R1 F60;`	
	(Macro call with flange-1 parameters)
`GOTO 6;`	

```
N2 G65 P8021 X#101 Y#102 Z10 D100 H6 S#103 R1 F50;
                    (Macro call with flange-2 parameters)
GOTO 6;
N3 G65 P8021 X#101 Y#102 Z15 D150 H8 S#103 R1.5 F40;
                    (Macro call with flange-3 parameters)
GOTO 6;
N4 G65 P8021 X#101 Y#102 Z20 D200 H10 S#103 R2 F30;
                    (Macro call with flange-4 parameters)
GOTO 6;
N5 G65 P8021 X#101 Y#102 Z20 D300 H12 S#103 R2 F30;
                    (Macro call with flange-5 parameters)
N6 M09;
M30;
```

The readers should observe the programming styles adopted in O8020 and O8021. O8020 includes most of the required set-up-related commands (such as tool change), requiring very few of these in the calling program. On the other hand, O8021 has none of these commands, which requires that these be inserted in the calling program. The first style is considered better, provided the macro is made flexible enough by using local variables for controlling its action. For example, a limitation of O8020 is that it will always use T0101 for roughing and T0303 for finishing. If some other tool/offset number is desired to be used, the macro would need to be modified, which is not considered good practice. Usually, macros are written by expert programmers and are edit-protected through parameters. Hence, local variables for roughing and finishing tool numbers should have been added in the argument list. On the other hand, even though several set-up commands are needed for calling O8021, this macro can be used in different cutting conditions, without any modification in it. Every approach has its own advantages and limitations. It is, basically, a matter of individual choice and specific requirements.

CHAPTER 10

Custom Canned Cycles

10.1 Introduction

It has already been discussed in Sec. 7.3 (subsection Call with User-Defined G-Code) how a new G-code can be defined or an existing one be redefined. The basic aim of defining a new G-code, instead of using a G65/G66 macro call, is that one need not know anything about macro programming to be able to call a macro. The defined G-code would call a macro without using G65/G66. It can be used the way a predefined canned cycle such as G71 on a lathe is used, with the difference that the used-defined G-code would always be a single-block code. In fact, all built-in canned cycles internally call some pre-defined macros only, without we realizing this. One only needs to know the significance of different letter addresses in such codes. The new G-code also can be used in the same manner. Thus, in a way, it enhances the standard control features, by making available some extra G-codes for some complex machining applications. As many as 10 new G-codes can be defined (including redefined G-codes).

There might be situations where no standard G-code would be quite suitable. A deep-hole drilling, requiring peck drilling with progressively reducing peck lengths, is one such example on a milling machine. Similarly, on a lathe, G74 is available for peck drilling, but the tool does not retract to outside the hole, after a peck. This might make deep-hole drilling difficult. Moreover, some of the G-codes, such as helical interpolation, are available as options, which need to be additionally purchased. In all such cases, it may be possible to define/redefinea G-code to suit particular application. Two examples, one for a lathe and one for a milling machine, are discussed next.

10.2 Deep-Hole Peck Drilling on a Lathe

The pecking action of the standard G74 cycle on a lathe is shown in Fig. 10.1. G74 is designed for drilling deep holes, but it does not work

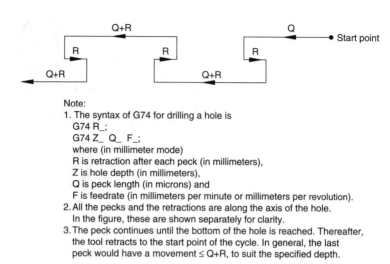

Note:
1. The syntax of G74 for drilling a hole is
   G74 R_;
   G74 Z_ Q_ F_;
   where (in millimeter mode)
   R is retraction after each peck (in millimeters),
   Z is hole depth (in millimeters),
   Q is peck length (in microns) and
   F is feedrate (in millimeters per minute or millimeters per revolution).
2. All the pecks and the retractions are along the axis of the hole.
   In the figure, these are shown separately for clarity.
3. The peck continues until the bottom of the hole is reached. Thereafter,
   the tool retracts to the start point of the cycle. In general, the last
   peck would have a movement ≤ Q+R, to suit the specified depth.

FIGURE 10.1　Peck-drilling cycle (G74) on a lathe.

satisfactorily when the hole is too deep, because though it does break the chips, it does not clear them. While this would not cause any problem with a *through-coolant* drill, it is likely to result in a reduced tool life because of chip clogging and coolant starvation, if solid drills are used. G74 is, in fact, the lathe equivalent of G73 on a milling machine, with similar limitations. While G83 is available on a milling machine, which clears the chips by retracting to outside the hole (up to R-point) after each peck, it is usually available on a lathe only if it has live tooling. Therefore, it is desirable to have a user-defined G-code on a lathe, which retracts the tool to outside the hole after each peck. The desired toolpath is shown in Fig. 10.2, where the G174 code number has been especially chosen to indicate that it is a modified version of G74. In fact, it is for this very reason that the peck length in G174 has been kept in microns, not in millimeters. This ensures that the Z-, Q- and F-words of both the cycles have the same meanings. Thus, both G74 and G174 can be used in the same manner, except that G174 would have a "missing" first block. Any programmer would be able to make use of G174, as if it were a built-in canned cycle of the control, without bothering about how it was made available on the machine. And so what if the programmer had never heard about macro programming!

Our next task is to define such a G-code. Referring to Table 7.1, 174 would need to be stored in parameter number 6051 (say), while the corresponding macro O9011 should generate the desired toolpath, as shown in Fig. 10.2, for which the following algorithm can be used:

1. Store the initial Z-position.

2. Calculate <required depth>. (Drilling to be started from the initial Z-position.)

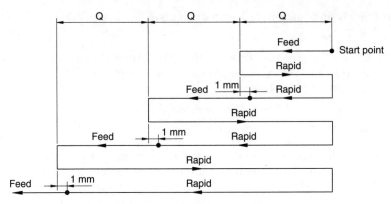

Note:
1. The syntax of G174, a user-defined cycle, for drilling a deep hole can be
   G174 Z_ Q_ F_;
   where (in millimeter mode)
   Z is hole depth (in millimeters),
   Q is peck length (in microns) and
   F is feedrate (in millimeters per minute or millimeters per revolution).
2. All the pecks and the retractions are along the axis of the hole. In the figure,
   these are shown separately for clarity.
3. The designed toolpath is same as that in the standard G74 cycle, except that the retraction
   at the end of each peck is up to the start point, for driving the chips to outside the hole.
4. All the outward motions (retractions) are rapid motions.
5. Each inward motion is rapid up to 1 mm before the previous drilled depth, after which feed
   motion starts, to increase the depth by the specified peck length.
6. The peck continues until the bottom of the hole is reached. Thereafter, the tool retracts to
   the start point of the cycle. The last peck would have an adjusted movement, to suit the
   specified depth.

FIGURE **10.2**   Deep-hole peck drilling on a lathe.

3. If <required depth> is less than or equal to Q, go to step 15.

4. Drill at the specified feedrate up to Q depth from the initial position.

5. Store the current Z-position.

6. Set <required depth> = <required depth> – Q

7. Retract at rapid rate to the initial position.

8. If <required depth> is less than or equal to Q, go to step 17.

9. Rapid motion up to 1 mm before the Z-position at the end of the previous peck.

10. Feed motion to increase the existing depth of the hole by Q.

11. Store the current Z-position.

12. Retract at rapid rate to the initial position.

13. Set <required depth> = <required depth> – Q

14. Go to step 8.

15. Single and continuous feed motion up to the final depth.

16. Go to step 18.

17. Drill the required depth, with rapid motion up to 1 mm before the previous drilled depth, followed by feed motion to reach the bottom of the hole.

18. Retract at rapid rate to the initial position.

19. Return to the calling program.

Macro O9011, which is based on the given algorithm, also includes some error traps in the beginning of the program, to detect certain incorrect ways of calling it. This macro would drill the hole using the spindle speed used in the calling program, by the existing tool in the spindle, as is done by G74. The macro is, however, designed to be called with a desired feedrate. If a feedrate is not commanded while calling the macro, the feedrate active in the calling program, at the time of calling the macro, would be used for drilling. The depth of hole can be commanded in both absolute form (Z-word) and incremental form (W-word), though simultaneous specification of the two is not permitted (which would alarm out). Finally, the macro does not require that the Z-datum necessarily be at the right face of the workpiece. The axial position of the desired hole is not at all restricted by Z-datum. The entire hole may lie to the right or to the left of the Z-datum, or it may even have Z-datum somewhere inside it.

The letter addresses in the macro-calling block (the corresponding local variables, as per argument specification 1, are given in brackets), and the common variables used in this macro are

Z (#26) = absolute Z-coordinate of the bottom of the hole

W (#23) = incremental Z-coordinate of the bottom of the hole (measured from the initial tool position)

Q (#17) = peck length (in microns, in millimeter mode)

F (#9) = feedrate

#100 = stores Z-coordinate of the initial tool position

#101 = stores <required depth> (the remaining depth to be drilled), which is updated after every peck

#102 = stores Z-coordinate at the end of a peck, hence this also is updated after every peck

```
O9011 (DEEP HOLE DRILLING ON LATHE);
IF [#17 EQ #0] THEN #3000 = 1 (PECK LENGTH NOT
SPECIFIED);
IF [[#23 EQ #0] AND [#26 EQ #0]] THEN #3000 = 2 (DEPTH NOT
SPECIFIED);
IF [[#23 NE #0] AND [#26 NE #0]] THEN #3000 = 3 (BOTH Z
AND W SPECIFIED);
```

```
IF [[#26 NE #0] AND [#5042 LE #26]] THEN #3000 = 4
(INCORRECT INITIAL POSITION);
IF [#23 GT 0] THEN #3000 = 5 (W MUST BE NEGATIVE);
IF [[#26 EQ #0] AND [#23 EQ 0]] THEN #3000 = 6 (ZERO DEPTH
SPECIFIED);
IF [#9 EQ #0] THEN #9 = #4109;
IF [#9 EQ 0] THEN #3000 = 7 (FEEDRATE NOT SPECIFIED);
```
(Alarms out if incorrect inputs are specified)

```
#100 = #5042;
```
(Initial Z-position stored. **Step 1 complete**)

```
#101 = #5042 - #26;
```
(Calculates required depth when Z is specified)

```
IF [#23 NE #0] THEN #101 = ABS[#23];
```
(Calculates required depth when W is specified. **Step 2 complete**)

```
#17 = ABS[#17] / 1000;
```
(Peck length in microns converted to millimeters. Both positive and negative values for peck length allowed. Pecking would always be done in the negative Z-direction)

```
IF [#101 LE #17] GOTO 10;
```
(If the required depth is less than or equal to the specified peck length, jump to block N10 to drill the hole without pecking. **Step 3 complete**)

```
G01 W-#17 F#9;
```
(Drilling done up to the specified peck length, measured from the initial Z-position. **Step 4 complete**)

```
#102 = #5042;
```
(Current Z-position stored. **Step 5 complete**)

```
#101 = #101 - #17;
```
(Required depth updated. **Step 6 complete**)

```
G00 Z#100;
```
(Rapid retraction to the initial Z-position. **Step 7 complete**)

```
WHILE [#101 GT #17] DO 1;
```
(Drilling in a loop. If the required depth becomes smaller than or equal to the specified peck length, exit from the loop. **Step 8 complete**)

```
G00 Z[#102 + 1];
```
(Rapid motion up to 1 mm to the right of the Z-position at the end of the previous peck. **Step 9 complete**)

```
G01 Z[#102 - #17];
```
(Feed motion to increase the depth by the specified peck length. **Step 10 complete**)

```
#102 = #5042;
```
(Current Z-position stored. **Step 11 complete**)

```
G00 Z#100;
```
(Rapid retraction to the initial Z-position. **Step 12 complete**)

```
#101 = #101 - #17;
```
(Required depth updated. **Step 13 complete**)

END 1;	(Jump to the WHILE block to check the loop condition again. **Step 14 complete**)
GOTO 20;	(Jump to N20 to drill the remaining depth)
N10 G01 W-#101 F#9;	(Single feed motion to reach the final depth. **Step 15 complete**)
GOTO 30;	(Jump to N30 for rapid retraction, and return to the calling program. **Step 16 complete**)
N20 G00 Z[#102 + 1];	(Rapid motion up to 1 mm before the previous drilled depth)
G01 Z[#102 - #101];	(Remaining depth drilled. **Step 17 complete**)
N30 G00 Z#100;	(Rapid retraction to the initial Z-position. **Step 18 complete**)
M99;	(Return to the calling program. **Step 19 complete**)

This macro has been designed to be used in millimeter mode. However, it is very easy to modify it to make it suitable for both millimeter mode (G21) and inch mode (G20). In inch mode, the 1 mm clearance amount (refer to Fig. 10.2) would need to be replaced by 0.04 in. For this, replace the clearance amount (1) by a new variable, say, #103, everywhere in the macro (there are two such occurrences). Read system variable #4006 to find out whether G20 or G21 is active. If #4006 contains 20, set #103 to 0.04, and if #4006 contains 21, set #103 to 1, in the beginning of the macro. In inch mode, the divisor 1000 also would need to be replaced by 10,000, in the block that redefines #17. The required modification in the program has been left as an exercise for the readers.

Structurewise, macro O9011 is no different from other macros. It can be called, in the usual manner of calling a macro, by

```
G65 P9011 Z_ / W_ Q_ F_; or
G65 P9011 Z_ / W_ Q_;
```

However, it also defines G174 if 174 is stored in parameter 6051, as already mentioned. In such a case, G174 ... would be equivalent to G65 P9011 ...

To illustrate the use of the new G-code (G174), consider the part shown in Fig. 10.3. For machining it with G174, just prepare a program for G74, and replace G74 by G174 (how simple it is!). The corresponding block is shown boldfaced, for easy identification:

```
G21 G97 G98;
G28 U0;
G28 W0;
T0101;
```

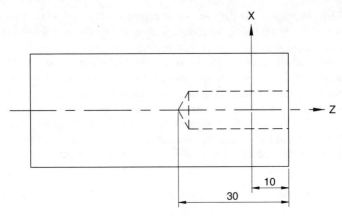

Note:
The usual choice for the Z-datum is at the right face of the workpiece. It is shown at an arbitrary place only to illustrate the use of macro O9011.

Figure 10.3 A cylindrical part with a hole along its axis.

```
G00 Z12;
X0;
M03 S1000;
M08;
G174 Z−20 Q5000 F20;
(or, G174 W−32 Q5000 F20);
M09;
M05;
G28 U0;
G28 W0;
M30;
```

Note that the conical shape at the bottom of the hole is caused by the tip angle of the drill bit. Using trigonometric relations, it can be shown that, for a given tool diameter D and tip angle θ, the height h of the cone is given by

$$h = \frac{D}{2} \times \tan\left(90 - \frac{\theta}{2}\right)$$

For the usual tip angle of 119°, this equation gets approximately simplified to

$$h = 0.3D$$

So, to obtain a cylindrical hole of a certain length, with, say, a 10-mm-diameter drill bit, 3 mm should be added to the desired cylindrical length. Hence, in Fig. 10.3, the length of the cylindrical portion would be about 27 mm, if the tool diameter is 10 mm.

This completes the discussion on how to define a new G-code on a lathe. It is also possible to redefine an existing G-code, to add certain desirable feature(s) to it. For example, commanding G01 with very low spindle speed is likely to be a programming mistake. (How can you machine unless the spindle is rotating at a reasonable speed?) Therefore, it may be desirable that commanding G01 with a very low spindle speed pauses the program execution, and waits for a confirmation from the operator (by pressing CYCLE START again), to continue the execution. Similarly, commanding both X and U (or Z and W) in the same block is meaningless, and should alarm out. The required modification in G01 can be done in the following manner:

```
IF [[#24 NE #0] AND [#21 NE #0]] THEN 3000 = 1 (BOTH X
AND U SPECIFIED);
IF [[#26 NE #0] AND [#23 NE #0]] THEN 3000 = 2 (BOTH Z
AND W SPECIFIED);
IF [#9 EQ #0] THEN #9 = #4109;
IF [#19 EQ #0] THEN #19 = #4119;
IF [[#19 LT 100] AND [#4002 EQ 97]] THEN #3006 = 1 (IS
LOW RPM OK);
IF [[#19 LT 10] AND [#4002 EQ 96]] THEN #3006 = 1 (IS
LOW CSS OK);
G01 X#24 U#21 Z#26 W#23 F#9 S#19;
```

It is left as an exercise for the readers to interpret the given program blocks and complete this exercise by editing a parameter among 6050–6059 (storing 1 in it) and defining the corresponding program number among O9010–O9019.

Recall that a G-code, inside a macro called by the same or a different G-code (other than G65/G66), is treated as the predefined standard G-code, with the usual function. Hence, G01 inside the macro called by G01 would be considered linear interpolation only; it would not call the same macro again, forming an endless loop.

Note that the chosen minimum values for rpm and CSS (100 and 10 m/min, respectively) are arbitrary. Specify suitable values for specific applications. Moreover, just checking the S-value does not confirm that M03 or M04 has been commanded to start the spindle! In fact, there is no way to determine whether or not the spindle is rotating through a program. The system variable for the M-code (#4113) stores the **last** commanded M-code number. For example, if M08 is commanded after M03, #4113 would contain 8. Therefore, to be on the safer side, one may tend to command M03 or M04 in the macro, but this also is not foolproof because only one of the two would be suitable in a particular machine set-up (whether front- or rear-type lathe, and normal or inverted tool clamping), which the macro programmer can only guess. Hence, the discussion on verification of spindle speed in the macro, practically remains only theoretical, though it is a good programming exercise. One should know the capabilities as well as limitations of macro programming. Additionally, try to make the macro suitable for inch mode as well.

10.3   Drilling with Reducing Peck Lengths on a Milling Machine

G73 and G83 are the two peck-drilling cycles available on milling machines. G73 is a high-speed peck-drilling cycle, using a very small retraction after each peck. It is used when the hole is not too deep. G83 retracts the tool to R-point after every peck. This clears the chips completely, making the cycle suitable for deep holes. If, however, the hole is too deep, coolant starvation becomes an issue after a few pecks. The only solution is to use pecks of gradually reducing lengths, by a desired factor. Such a peck-drilling cycle is also referred to as a regressive peck-drilling cycle. Unfortunately, Fanuc control does not provide a built-in regressive peck-drilling cycle. Therefore, in case it is really needed, one would need to write a macro for it. Better still, a new G-code (say, G183 that indicates a modified G83) may be defined for it. For this, 183 would need to be stored in, say, parameter 6052, while the corresponding program O9012 should generate the desired toolpath, as shown in Fig. 10.4.

The algorithm for macro O9012 would generally be similar to the one used for macro O9011, with the main difference that the peck length would need to be calculated before each peck:

1. Store the initial Z-position.

2. Rapid motion to the XY-position (the hole center).

3. Rapid motion to R-point.

4. Calculate <required depth> (drilling to be started from the R-point).

5. If <required depth> is less than or equal to Q, go to step 20.

6. Drill at the specified feedrate up to Q depth from the R-point.

7. Store the current Z-position.

8. Set <required depth> = <required depth> − Q.

9. Retract at rapid rate to the R-point.

10. If <required depth> is less than or equal to M or the new peck length (which is equal to the previous peck length multiplied by regression factor I), go to step 18.

11. Calculate the new peck length. If it comes out to be smaller than M, set it equal to M.

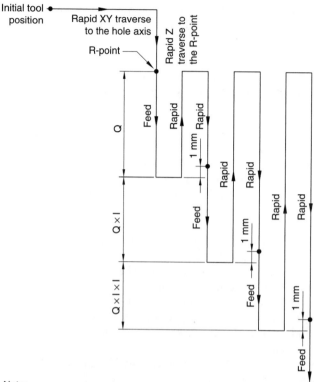

Note:

1. The syntax of G183, a regressive peck-drilling cycle on a milling machine, can be
   G183 X_ Y_ Z_ R_ Q_ I_ M_ F_;
   where
   X, Y is hole location,
   Z is the hole depth,
   R is the R-point level,
   Q is the first peck length,
   I is the regression factor,
   M is the minimum peck length and
   F is the feedrate.
2. The tool first moves to the XY-point (hole axis), and then to the R-point, at rapid traverse r
   after which drilling starts. The designed toolpath is the same as that in the standard G83
   cycle, except that the subsequent peck lengths get reduced by the regression factor
   $(Q \rightarrow Q \times I \rightarrow Q \times I \times I,$ and so on).
3. All the pecks and the retractions are along the axis of the hole. In the figure, these are
   shown separately for clarity.
4. If a calculated peck length comes out to be smaller than the specified minimum peck
   length, this as well as all the subsequent peck lengths are clamped to the specified
   minimum value.
5. All the outward motions (retractions) are rapid motions.
6. Each drilling motion is rapid up to 1 mm before the previous drilled depth, after which
   feed motion starts, to increase the depth by the calculated peck length.
7. The peck continues until the bottom of the hole is reached. Thereafter, the tool retracts to
   R-point or the initial Z-level. The last peck would have an adjusted movement, to suit the
   specified depth.

Figure 10.4 Regressive peck-drilling cycle on a milling machine.

12. Rapid motion up to 1 mm above the Z-position at the end of the previous peck.

13. Feed motion to increase the existing depth of hole by the new peck length.

14. Store the current Z-position.

15. Retract at rapid rate to R-point.

16. Set <required depth> = <required depth> – <current peck length>.

17. Go to step 10.

18. Drill the required depth, with rapid motion up to 1 mm above the previous drilled depth, followed by feed motion to reach the bottom of the hole.

19. Go to step 21.

20. Single and continuous feed motion up to the final depth.

21. Retract at rapid rate to the R-point if G99 is active, or the initial Z-position if G98 is active.

22. Return to the calling program.

The local and common variables used in this macro are given below (all coordinates are in the absolute coordinate mode, G90):

X (#24) = X-coordinate of the hole center

Y (#25) = Y-coordinate of the hole center

Z (#26) = Z-coordinate of the bottom of the hole

R (#18) = Z-coordinate of R-point

Q (#17) = first peck length

I (#4) = regression factor (such that any subsequent peck length is equal to the previous peck length multiplied by the regression factor. Hence, I = 1 corresponds to constant peck lengths)

M (#13) = minimum peck length

F (#9) = feedrate

#100 = stores Z-coordinate of the initial tool position

#101 = stores <required depth> (the remaining depth to be drilled), which is updated after every peck

#102 = stores Z-coordinate at the end of a peck, hence this also is updated after every peck

#103 = current peck length (calculated as its previous value multiplied by #4)

```
O9012 (REGRESSIVE PECKING ON MILL M/C);
IF  [#24  EQ  #0]  THEN  #3000  =  1  (SPECIFY  HOLE
X-COORDINATE);
```

IF [#25 EQ #0] THEN #3000 = 2 (SPECIFY HOLE Y-COORDINATE);

IF [#26 EQ #0] THEN #3000 = 3 (SPECIFY HOLE DEPTH);

IF [#18 EQ #0] THEN #3000 = 4 (SPECIFY R-POINT);

IF [#17 EQ #0] THEN #3000 = 5 (SPECIFY FIRST PECK LENGTH);

IF [#13 EQ #0] THEN #3000 = 6 (SPECIFY MINMUM PECK LENGTH);

IF [#4 EQ #0] THEN #3000 = 7 (SPECIFY REGRESSION FACTOR);

IF [#9 EQ #0] THEN #9 = #4109;

IF [#9 EQ 0] THEN #3000 = 8 (SPECIFY FEEDRATE);

	(Alarms out if inputs are missing)
#100 = #5043;	(Initial Z-position stored. **Step 1 complete**)
G00 X#24 Y#25;	(Rapid traverse to the hole center. **Step 2 complete**)
Z#18;	(Rapid traverse to the R-point. **Step 3 complete**)
#101 = #18 − #26;	(Calculates required depth. **Step 4 complete**)

IF [#101 LE 0] THEN #3000 = 9 (IMPROPER R OR Z SPECIFIED);

#17 = ABS[#17];	(Both positive and negative values for the first peck length allowed. Pecking would always be done in the negative Z-direction)
#13 = ABS[#13];	(Both positive and negative values for the minimum peck length allowed)
IF [#101 LE #17] GOTO 10;	(If the required depth is less than or equal to the first peck length, jump to block N10 to drill the hole without pecking. **Step 5 complete**)
G01 Z[#18 − #17] F#9;	(Drilling done up to the first peck length, measured from the R-point. **Step 6 complete**)
#102 = #5043;	(Current Z-position stored. **Step 7 complete**)
#101 = #101 − #17;	(Required depth updated. **Step 8 complete**)
G00 Z#18;	(Rapid retraction to R-point. **Step 9 complete**)
#103 = #17;	(Current peck length initialized)
WHILE [[#101 GT #13] AND [#101 GT [#103 * #4]]] DO 1;	(Drilling in a loop. If the required depth becomes smaller than or equal to the minimum peck length or the new peck length, exit from the loop. **Step 10 complete**)

```
#103 = #103 * #4;
IF [#103 LT #13] THEN #103 = #13;
```
                                    (New peck length calculated.
                                    Step 11 complete)

```
G00 Z[#102 + 1];
```
                                    (Rapid motion up to 1 mm above
                                    the Z-position at the end of the pre-
                                    vious peck. **Step 12 complete**)

```
G01 Z[#102 - #103];
```
                                    (Feed motion to increase the depth
                                    by the calculated peck length.
                                    Step 13 complete)

```
#102 = #5043;
```
                                    (Current Z-position stored. **Step 14
                                    complete**)

```
G00 Z#18;
```
                                    (Rapid retraction to R-point.
                                    Step 15 complete)

```
#101 = #101 - #103;
```
                                    (Required depth updated. **Step 16
                                    complete**)

```
END 1;
```
                                    (Jump to the WHILE block to check
                                    the loop condition again. **Step 17
                                    complete**)

```
G00 Z[#102 + 1];
G01 Z#26;
```
                                    (Remaining depth drilled. **Step 18
                                    complete**)

```
GOTO 20;
```
                                    (Jump to N20 for final retraction.
                                    Step 19 complete)

```
N10 G01 Z#26 F#9;
```
                                    (Single feed motion to reach the
                                    final depth. **Step 20 complete**)

```
N20 G00 Z#18;
IF [#4010 EQ 99] GOTO 30;
```
                                    (Retracts to R-point)
                                    (Jumps to program end if G99 is
                                    active)

```
G00 Z#100;
```
                                    (Retracts to the initial Z-level, if
                                    G98 is active. **Step 21 complete**)

```
N30 M99;
```
                                    (Return to the calling program.
                                    Step 22 complete)

The given macro uses millimeter mode, and can be called only in the absolute coordinate mode. Though it is easy to make it suitable for inch mode as well, it would require a little more effort to use it in incremental coordinate mode (G91). This is being left as an exercise for the readers.

To illustrate the use of this macro, consider the part shown in Fig. 10.5, where it is assumed that the holes are so deep (i.e., with a large length to diameter ratio) that these require the use of regressive peck drilling cycle (G183). The G183 blocks are shown highlighted for easy identification:

```
G21 G94 G54;
G91 G28 Z0;
G28 X0 Y0;
M06 T01;
G90 G00 X0 Y0;
```

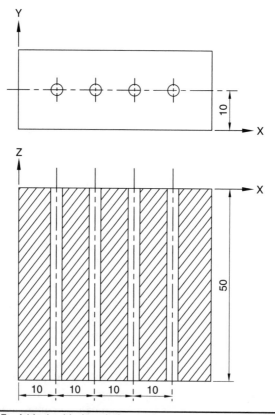

Figure 10.5 A block with deep holes.

```
G43 H01 Z100;
M03 S1000;
M08;
G99;
G183 X10 Y10 Z-53 R2 Q10 I0.9 M5 F20;
G183 X20 Y10 Z-53 R2 Q10 I0.9 M5 F20;
G183 X30 Y10 Z-53 R2 Q10 I0.9 M5 F20;
G98;
G183 X40 Y10 Z-53 R2 Q10 I0.9 M5 F20;
M09;
M05;
G91 G28 Z0;
M30;
```

Note that the first three holes have been made in G99 mode, and the last one in G98 mode. Hence, the tool would stay at the R-point (2 mm above the top surface of the block) after making each of the first three holes, whereas it would retract to the initial Z-level (100 mm above the block) after completing the last hole. Moreover, the last three calls may not have F-words, since the macro is designed to automatically use the previous feedrate, if a feedrate is not specified while calling the macro.

The macro call (with G65 or by other methods) should appear in a **separate** block, as in the given example where G98 and G99 are not clubbed with G183. If a G-code is commanded to the left of the macro call, in the same block, it is ignored by the control. On the other hand, a G-code to the right of macro call alarms out, because it is considered an argument of the macro, whereas a G-word is not allowed as an argument. M98 has a different behavior in this situation. It does not ignore the G-code or alarms out, but does not wait for the completion of G-code execution. Hence, if it is a movement code (such as G01), the subprogram execution would start while the tool is still moving. Therefore, to avoid any confusion, **subprograms and macros should always be called in separate blocks**.

Some of the canned cycles, as well as codes for special functions such as helical interpolation, are available as options, for which additional payment is required to be made to the control manufacturer. It is also possible to activate any option at a later date, provided the hardware of the machine tool is equipped to handle it (in case of doubt, consult the MTB). However, as the given examples demonstrate, it is possible for the users themselves to define the required G-codes, with desired features. And these can be made even better than the standard codes, in certain cases! If you know macro programming well enough, and are ready to flex your brain, you can do virtually anything.

CHAPTER 11

Probing

11.1   Introduction

The productivity of a CNC machine can be increased if the set-up time and inspection time are reduced, by automating these time-consuming processes. The technique broadly involves automatic measurement of certain dimensions, and taking machining-related decisions, based on the result of measurement. This process is referred to as *probing*.

The device that is commonly used for probing is a *touch probe* that remains electronically connected (through a cable or an infrared/radio signal) to the control. A touch probe senses very small displacement, and sends out an electrical signal (called *skip signal*) when "touched." The control immediately stores the probe/tool position the moment it receives the signal, in system variables, #5061–#5064 (refer to Table 3.11), and stops further movement of the tool. The stored positions at planned locations are used for calculating the desired dimensions.

There are two basic types of touch probes: *tool probe* and *work probe*. A tool probe is used for automating the tool offset setting procedure, which is a manual process involving error-prone human judgment. It is a fixed-type probe, and remains attached to a stationary part on the machine body. The tool, which is to be probed, is made to touch it (probe). Since the position of the probe is known, information regarding the geometry of the tool can be extracted. Moreover, a periodic repetition of this process gives information about the extent of tool wear, so that the wear offset could be suitably edited to nullify the error introduced due to tool wear. A sudden and large change in tool dimensions indicates tool breakage.

Work probe, on the other hand, is a moving-type probe. It is held by the machine the way a tool is held. It is moved like a tool, so as to touch the workpiece/finished part at desired locations. The stored positions at skip signals are used to calculate the desired dimensions of the workpiece/finished part. The examples in this chapter pertain to this type of probe only. Such a probe has a long stylus, with a small ruby ball at its end (ruby is used since it is hard and hence prevents any change in the dimension of the ball due to wearing). When the ball presses against a fixed surface, the stylus deflects a little bit, and disturbs an optical focusing arrangement inside the probe. A differential photocell senses the change in the position of the image point, and triggers a signal. The diameter of the ball has to be taken into account in calculations, if required. (In some cases it does not come into the picture, e.g., it may cancel out while taking a difference in two positions.)

The stylus of the probe is supported by a spring system, so that it does not break when suddenly deflected by a hard surface. Typically, a deflection of up to 5 mm at the probe tip (ball) is permitted. Note that the stored positions are not likely to be very accurate because of tip overshoot. One way to handle the overshoot is to use a fixed and controlled speed (feedrate), and use a calibration chart to estimate the overshoot. Another way involves probing at a very slow speed. Since it would take too much time, the surface is probed twice. First it is touched with a high speed, then, when the skip signal is triggered, it is backed off by a small amount (say, 2 to 3 mm), and moved again toward the surface with a very slow speed. This technique gives good results in a very short time, and hence is preferred most of the time. The examples in this chapter use the same method.

11.2   Skip Function on a CNC Machine

Apart from the required hardware (a touch probe and a physical connection for receiving the skip signal by the control), the control also should be equipped with the skip-function feature, at software level. The associated G-code is G31. Hence, unless G31 (or a similar code on controls other than Fanuc) is available, probing would not be possible. G31 might be an optional feature on some control versions.

G31 is very much similar to G01, except that when the control receives a skip signal, the execution of G31 immediately terminates, the tool/probe position at that instant gets automatically stored in certain system variables, and the execution of the next block starts. When we talk about moving the tool/probe toward the surface to be probed, it actually refers to movement with G31 command. Figure 11.1 explains the function of G31, with and without a skip signal, on a milling machine. Without a skip signal, the movement occurs up to the programmed end point, whereas the movement gets terminated and the execution of the next block starts, the moment control receives a skip signal.

11.3   Probing a Pocket

Measuring the dimensions of a pocket is a very simple example of probing, but it does explain the underlying principles very clearly. The probe simply has to touch the top surface of the part and the bottom of the pocket, one by one, for depth measurement. The difference

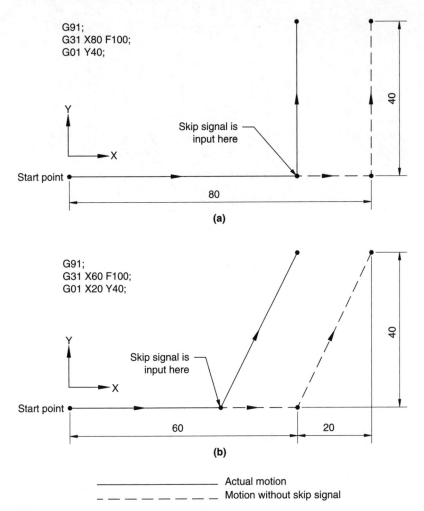

G91;
G31 X80 F100;
G01 Y40;

Y

X

Skip signal is input here

Start point

80

40

(a)

G91;
G31 X60 F100;
G01 X20 Y40;

Y

X

Skip signal is input here

Start point

60   20

40

(b)

_____ Actual motion
_ _ _ _ _ _ _ _ Motion without skip signal

Note:
G31 is designed to work with an obstacle in the toolpath, so as to generate
a skip signal for the purpose of measurement. It has no other use. So, an absence
of skip signal before reaching the specified end point in G31 quite possibly
indicates an error condition, either in programming or due to unexpected deviation in
part dimensions. Hence, some controls issue an alarm, and terminate further execution
of the program, in such cases.

FIGURE 11.1 Motion associated with the skip function (G31).

in the Z-positions at respective skip signals would be the depth of the
pocket. This is shown in Fig. 11.2 (positions B and D, respectively).
Note that the diameter of the ball does not come into the picture
because the same point of the ball touches the two surfaces. Similarly, other dimensions (say, the length/width of the pocket and the

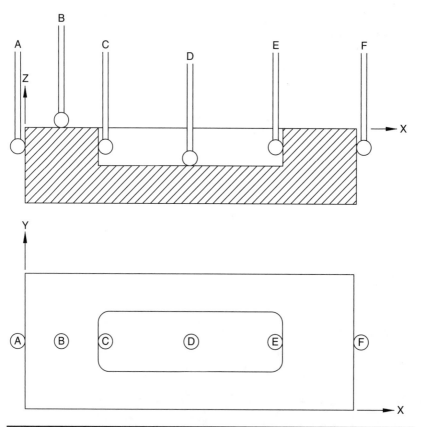

Figure 11.2   Probing a pocket on a milling machine.

length/width of the part) can be measured. The length of the pocket can be measured by touching the pocket at, say, positions C and E, and then taking the difference in the stored X-positions. As is clear from the geometry, the diameter of the ball would need to be added to the obtained length. On the other hand, when the length of the part is measured by touching it at positions A and F, the diameter of the ball would need to be subtracted from the obtained value.

The dimensions of the pocket are not important for explaining the probing principle. However, for the purpose of writing a program, it is assumed that the "expected" size of the block and the pocket are 90 mm × 36 mm × 20 mm and 50 mm × 16 mm × 10 mm, respectively, with their centers at (45, 18). It is also assumed that the probe occupies the position of tool post number 1, with 01 offset number for length compensation. The diameter of the ball is assumed to be 5 mm. The program for this problem is intentionally not made for a general case since its only aim is to explain the probing principle, without involving additional complications:

```
O0020;
G21 G94;
G91 G28 Z0;
G28 X0 Y0;
M06 T01;
G90 G00 X10 Y18;
G43 H01 Z50;
G91 G31 Z-100 F200;
```
(The probe approaches position B. It is expected that the probe would definitely touch the surface within the specified distance. The distance 100 has no special significance, except that it must be larger than the gap between the probe and the surface to be probed. Note that this as well as all subsequent motions are in incremental mode)

```
Z2;
```
(Backs off by 2 mm. G00 is implied. It is assumed that the overshoot is less than 2 mm)

```
G31 Z-5 F10;
```
(Very low approach speed, to ensure accuracy)

```
#100 = #5063;
```
(Z-coordinate at the skip signal stored, which corresponds to position B)

```
Z5;
X35;
```
(Probe positioned at the center of the pocket)

```
G31 Z-100 F200;
```
(Probe approaches the bottom of the pocket)

```
Z2;
G31 Z-5 F10;
#101 = #5063;
```
(Z-coordinate corresponding to position D stored)

```
#500 = #100 - #101;
```
(Depth of the pocket calculated. Diameter of the probe does not come into the picture)

```
Z3;
G31 X-100 F200;
```
(Approaches the left side of the pocket)

```
X2;
G31 X-5 F10;
#102 = #5061;
```
(X-coordinate at position C stored)

```
G31 X100 F200;
```
(Approaches the right side of the pocket)

```
X-2;
G31 X5 F10;
#103 = #5061;
```
(X-coordinate at position E stored)

```
#501 = #103 - #102 + 5;
```
(Length of the pocket calculated, in which diameter of the ball has been taken into account)

```
X-5;
```

```
Z10;
X50;
Z-10;
G31 X-100 F200;          (Approaches the right side of the
                          block)
X2;
G31 X-5 F10;
#104 = #5061;            (X-coordinate at position F stored)
X5;
Z10;
X-120;
Z-10;
G31 X100 F200;           (Approaches the left side of the block)
X-2;
G31 X5 F10;
#105 = #5061;            (X-coordinate of position A stored)
#502 = #104 - #105 - 5;  (Length of the block calculated, in
                          which diameter of the ball has been
                          taken into account)
X-5;
G28 Z0;
G28 X0 Y0;
M30;
```

The Y-dimensions of the pocket and the block can be determined in a similar manner. This has been left as an exercise for the readers. Probing on a lathe is done in a similar manner, where the probe is generally held like an internal tool.

11.4 Finding Center of a Hole

Sometimes the workpiece is a cast part with a preexisting hole, and the part program is written with workpiece zero point at the center of the hole. This requires that the spindle be exactly aligned with the axis of the hole, for determining work offsets along the X- and the Y-axis. This is typically done with the help of a spindle-type dial indicator. The end of the lever is made to touch the hole at some point, and the spindle is given a very low rpm. The X- and/or Y-positions are adjusted so as to get the same dial reading in all angular positions, which locates the center of the hole. Though this method is theoretically correct, and is used quite often, it involves trial and error. As a result, the time taken in datum setting largely depends on the skill of the operator. We will now see how this can be done automatically, using a touch probe.

Consider the hole shown in Fig.11.3. The principle involved in locating the center is very simple. The center would lie on the point of intersection of the perpendicular bisectors on any two chords. As a

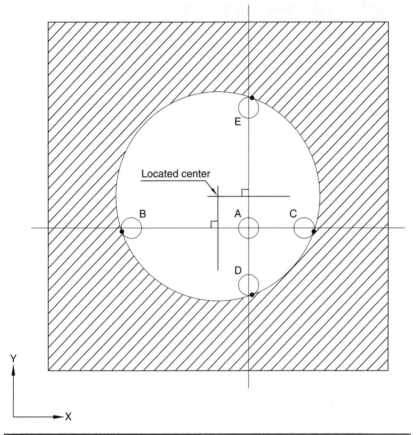

Figure 11.3 Locating center of a hole.

special case, if the two chords are chosen to be parallel to the X- and Y-axes, respectively, then the X-coordinate of the center would be same as that of the mid-point of the chord that is parallel to the X-axis. Similarly, the Y-coordinate of the center would be same as that of the mid-point of the chord that is parallel to the Y-axis. When the end-points of the chords are found using a touch probe, its diameter comes into the picture, because the circle is not touched by the same point on the ball. (The points on the ball, which touch the circle, are shown by dots in Fig. 11.3.) This, however, does not affect our calculations because of symmetry, as can be observed in the figure.

A macro, based on this principle, will be developed now. To locate the center of a hole, one simply has to place the ball of the probe somewhere inside the hole (say, position A), and then call the macro. After locating the center, the macro would place the workpiece zero point at that point, by editing the associated system variables.

```
O8022 (LOCATING HOLE CENTER ON MILL);
G21 G94;
#100 = #5041;
```
(X-coordinate at the initial position A stored. Refer to Table 3.11)

```
G91 G31 X-200 F200;
```
(Approaches position B)

```
G00 X2;
G31 X-5 F10;
#101 = #5061;
```
(X-coordinate of the center of the ball stored, at position B)

```
G31 X200 F200;
```
(Approaches position C)

```
X-2;
G31 X5 F10;
#102 = #5061;
```
(X-coordinate of the center of the ball stored, at position C)

```
#500 = [[#101 + #102] / 2];
```
(X-coordinate of the center calculated)

```
G90 X#100;
```
(Probe moves to position A)

```
G91 G31 Y-200 F200;
```
(Approaches position D)

```
G00 Y2;
G31 Y-5 F10;
#103 = #5062;
```
(Y-coordinate of the center of the ball stored, at position D)

```
G31 Y200 F200;
```
(Approaches position E)

```
Y-2;
G31 Y5 F10;
#104 = #5062;
```
(Y-coordinate of the center of the ball stored, at position E)

```
#501 = [[#103 + #104] / 2];
```
(Y-coordinate of the center calculated)

```
G90 X#500 Y#501;
```
(Probe moves to the center of the hole)

```
G65 P8008;
```
(Program O8008, which is given in Chap. 5, shifts the current workpiece zero point to the current tool position, on a milling machine. Just replace M30 by M99 in the last block of O8008, for using it as a macro/subprogram)

```
M30;
```

This macro, however, would not work if additional WCS, such as G54.1 P1, is currently active, or when the diameter of the hole is more than 200 mm. It is left as an exercise for the readers to make this macro suitable for these cases. Additionally, they may also try to calculate the radius of the hole. For this, calculate the distance between the center of the hole and any of the B, C, D, or E positions, using the coordinate-geometry formula, and add to it the radius of the ball. This would require the Y-coordinate of the initial tool position (at A), if positions B or C are selected for calculations. To avoid this, use positions D or E. The center of a rectangular pocket can be determined in a similar manner.

11.5   Finding Angle of an Edge

Sometimes the machining program for a rectangular block is written with program zero point at, say, the lower left corner of the block, with the coordinate axes being parallel to its edges. In such a case, a very accurate (hence expensive) fixture is needed, to hold the workpiece properly. An inaccurate fixture would hold it incorrectly, as shown in Fig. 11.4, where the coordinate system used for programming (X-Y) does not coincide with the workpiece coordinate system (X_w-Y_w). A solution to this problem would be to shift the workpiece zero point to point O, and then execute the machining program using coordinate-system-rotation feature (G68). Calling macro O8008, after placing the tool at point O, would shift the coordinate system, as was done in the previous example. For commanding G68, the angle θ (in degrees) must be known. The algorithm would be

1. Find angle θ.
2. Find X- and Y-coordinates of point O.
3. Place the tool at point O (at a safe height from the workpiece).
4. Call O8008.

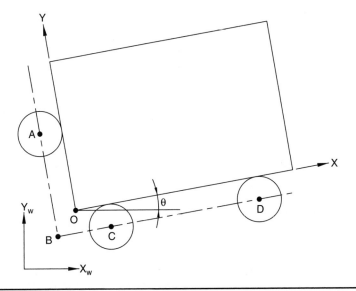

FIGURE 11.4   Finding the angle of an edge.

5. Command G17 G68 X0 Y0 R<θ> (X0 Y0, which is the center of rotation, can be omitted because the current tool position automatically becomes the center of rotation, if it is not specified).

6. Execute machining program.

7. Command G69 (cancel coordinate rotation).

8. End of program.

The X- and Y-coordinates of point O, and angle θ, can be determined using touch probes. For this, the probe would need to touch the orthogonal edges of the workpiece at three points, as shown in Fig. 11.4. Let the coordinates of probe center be (x_A, y_A), (x_C, y_C), and (x_D, y_D), respectively, corresponding to points A, C, and D. The equations given below can be used for necessary calculations. (Try to derive these, using coordinate geometry.)

The equations of lines CD and AB (which is orthogonal to CD) would be

$$y_{CD} = \frac{y_D - y_C}{x_D - x_C}(x - x_C) + y_C$$

$$y_{AB} = -\frac{x_D - x_C}{y_D - y_C}(x - x_A) + y_A$$

The coordinates (x_B, y_B) of the point of intersection of these lines (point B) can be found out by solving these equations simultaneously, which gives

$$x_B = \frac{m^2 x_C - m y_C + x_A + m y_A}{m^2 + 1}$$

$$y_B = \frac{-m x_C + y_C + m x_A + m^2 y_A}{m^2 + 1}$$

where

$$m = \frac{y_D - y_C}{x_D - x_C}$$

Angle θ would be given by

$$\theta = \tan^{-1} m$$

Finally, the coordinates of the program zero point O would be

$$x_O = x_B + \sqrt{2}\, r \cos(\theta + 45)$$

$$y_O = y_B + \sqrt{2}\, r \sin(\theta + 45)$$

where r is the radius of the ball of the probe.

The given algorithm, along with these equations, can be used for developing a macro for this problem. Set parameter 6004#0 to 1, for having the solution range of $\tan^{-1}$ (ATAN function) between $-180°$ and $180°$, though even if the range is $0°$ to $360°$ (with 6004#0 set to 0), it is mathematically equivalent. Therefore, do not worry about this parameter, if, say, $355°$ in place of the "expected" $-5°$ does not look awkward to you. It is left as an exercise for the readers to complete this problem.

11.6   Adjusting Wear Offset

As a final example, consider the case of getting oversized parts on a lathe, due to tool wear. Though this problem also can be solved by adjusting geometry offset, it is not a recommended method. The reason is that when the worn-out insert is replaced with a new insert, the original geometry offset would need to be used again. If the original values are lost, the time-taking procedure of setting geometry offset would have to be repeated. Therefore, the recommended practice is to start with zero wear offset for a new insert, and adjust it to take care of tool wear. When the insert is finally replaced with a new insert, the wear offset values are set to zero again. Geometry offset is never manipulated. This obviates the need for repeating the offset setting procedure.

When an external tool wears out, the external diameters on the part become larger than their programmed values. On the other hand, if an internal tool wears out, the internal diameters become smaller. This requires that, for a given programmed diameter, an external tool be brought nearer to the spindle axis, and an internal tool be moved away from it, by an amount equal to the observed error, to compensate for the wear. This can be done by manipulating the offset distances, which define the position of the coordinate system with respect to the fixed machine coordinate system. Since the control uses the algebraic sum of geometry and wear offsets (as well as all other offsets), the change in wear offset for an external tool would be negative, whereas it would be positive for an internal tool. Note that all the offset values are "diameter values," if diameter programming is being used.

The method of offset correction is the same for both internal and external tools (except for the opposite signs for the required change in wear offset values, and addition/subtraction of the ball diameter to/ from the observed diameter while measuring the part diameter).

Offset correction for an external tool is considered here, as an example. Since the amount of error remains same on all diameters, any convenient diameter can be chosen for determining the error. In the example given in Fig. 11.5, $\phi20$ diameter has been arbitrarily chosen for measurement, at Z–25 axial position.

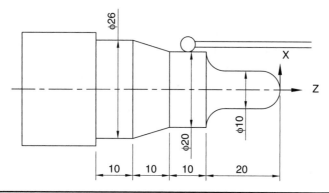

Figure 11.5 Diameter measurement on a lathe.

The algorithm for the macro O8023 is straightforward:

1. Determine the offset number of the current tool (which was used to machine the workpiece), using

   <current tool number> = FIX [#4120/100]

   <offset number of the current tool>
   = #4120 – <current tool number> × 100

2. Measure the specified diameter, at the specified location (X20 Z–25, in this example).

3. Subtract the specified diameter (20, in this case) from the obtained diameter, to determine the error due to tool wear.

4. Determine the system variable for wear offset corresponding to the offset number being used (refer to Tables 3.2 and 3.3).

5. Subtract the error from this system variable that contains the current value of wear offset corresponding to the offset number being used.

6. Return to the calling program.

```
O8023 (WEAR OFFSET CORRECTION ON LATHE);
M05;                         (It is expected that this macro would be
                             called with the spindle not rotating.
                             However, as a safety measure, M05 is
                             explicitly commanded here)
#100 = #4120;                (Current tool code stored)
#101 = FIX[#4120 / 100];     (Current tool number determined)
#102 = #4120 - #101 * 100;   (Offset number of the current tool deter-
                             mined. Step 1 complete)
G28 U0;                      (X-home)
```

`G28 W0;`	(Z-home)
`T#20;`	(The probe becomes the current "tool")
`G00 Z[#26 - #7 / 2];`	(Center of the probe at the specified Z-coordinate, where measurement is to be made. The radius of the probe has been subtracted from the specified value, assuming that the tip of the probe is its reference point)
`X[#24 + #7 + 4];`	(Probe at a clearance of 2 mm from the diameter to be measured. Make necessary changes in positioning toolpath, if an internal diameter machined by an internal tool is being probed. Moreover, both the plus signs would change to minus signs, for an internal measurement)
`#103 = #4005;`	(Stores 98 or 99, depending on which one of G98 and G99 is currently active)
`G98;`	(Since the spindle is stationary, this macro would not work in G99, the feed per revolution mode, because the feedrate would become zero with zero rpm)
`G31 U-10 F10;`	(Stops at the surface, where the diameter is to be measured)
`#104 = #5061;`	(X-coordinate of the center of the probe ball stored when it touches the part)
`#105 = #104 - #7;`	(Diameter of the part calculated. In the case of an internal diameter, the diameter of the ball would have to be added here. **Step 2 complete**)
`#106 = #105 - #24;`	(Error due to wear calculated. This quantity would be negative if probing is being done for an internal tool. **Step 3 complete**)
`#107 = 10000 + #102;`	(The system variable for wear offset, corresponding to the offset number being used, is determined. Replace 10,000 by 2000, if only 64 offset numbers are available. 10,000 series is for 99 offset numbers. **Step 4 complete**)
`#[#107] = #[#107] - #106;`	(Error subtracted from the current wear offset value. This expression is also correct also for internal measurement, since the sign of #106 would automatically become negative. **Step 5 complete**)
`G28 U0;`	(X-home)
`G28 W0;`	(Z-home)
`T#100;`	(Original tool called, with the updated wear offset value)
`G#103;`	(Modal G-code restored)
`M99;`	(Return to the calling program. **Step 6 complete**)

Wear offset correction for the Z-axis can be done in a similar manner, which is left as an exercise (usually, radial wear is more serious than the axial wear). One would need to touch a known Z-surface (which has been faced by the tool to be probed), and determine the error. The program for both internal and external tools would be the same in this case. The Z-axis wear offset system variables belong to #2100 series for 64 offsets, and #11000 series for 99 offsets (refer to Tables 3.2 and 3.3).

In order to use macro O8023, it would need to be called immediately after machining by the tool, for which wear adjustment is to be done. It is possible to continue machining further, with the **new** wear offset values. In fact, in the case of suspected inaccuracy, a single straight-turning command, with a small depth of cut, can be given first. The main machining codes can be given after calling this macro. This would automate the inspection and correction process. One may use this technique after, say, every 10 parts, which would ensure virtually zero rejection. The number of machined parts, in the current machining session, gets stored in system variable #3901 (refer to Sec. 3.5, subsection Number of Machined Parts). The main program may call/skip the probing macro by

```
<set-up commands>
<straight turning>
IF [#3901 LE 10] GOTO 100;
G65 P8023 T_ X_ Z_ D_;
#3901 = 0;
N100;
<machining program>
M30;
```

where the part count is reset to zero whenever the macro is called. It is left as an exercise for the readers to think how to implement the same logic if the part count variable #3901 is not to be reset (since it may be desired to keep track of the total number of the machined parts in the current machining session).

The arguments of macro call for O8023 pass data for the following:

T (#20) = four-digit tool code for the probe

X (#24) = diameter to be probed

Z (#26) = axis position where probing is to be done

D (#7) = diameter of the probe ball

Though one can develop macros for specific probing requirements, the probe manufacturers usually supply probes with software for common probing applications.

CHAPTER 12

Communication with External Devices

12.1 Introduction

Adequate description of how to control an external device through the PMC has already been given in Sec. 3.5 (subsection Interface Signals). Knowledge of Ladder language, however, is a prerequisite for programming the PMC. Additionally, one must also know input types (source/sink type), output types (source/sink/relay type), and wiring techniques for a PLC. Moreover, if noncontact-type proximity switches are desired to be used for sensing the presence of an object, one must know the characteristics of such sensors (source/sink type). The whole thing is pretty vast, and can be described in detail only in a text designed especially for this purpose. However, for the convenience of readers, a brief description is given in this chapter, the minimum one must know. The main purpose here is to refresh the memory, rather than explaining the fundamental concepts meticulously. Hence, a reader who is completely new to PLC applications would need to refer to some basic text on PLC.

12.2 Switching Principle

The input/output **digital** signals of a PLC are based on the principle of saturation-mode operation of a transistor, when it starts working like an electronic ON/OFF switch. (Before saturation, it serves the purpose of signal amplification.)

The representation of an NPN transistor is shown in Fig. 12.1, along with its functional model. I_B, I_C, and I_E are base current, collector current, and emitter current, respectively ($I_E = I_B + I_C$). The variable resistance between collector and emitter (R_{CE}), to which power source ($+V_{CC}$) and load are connected, can be thought of as being controlled by the base current; when the base current is zero, the resistance is infinite, which decreases with an increase in base current. As a result, when the base current increases, the collector current also increases.

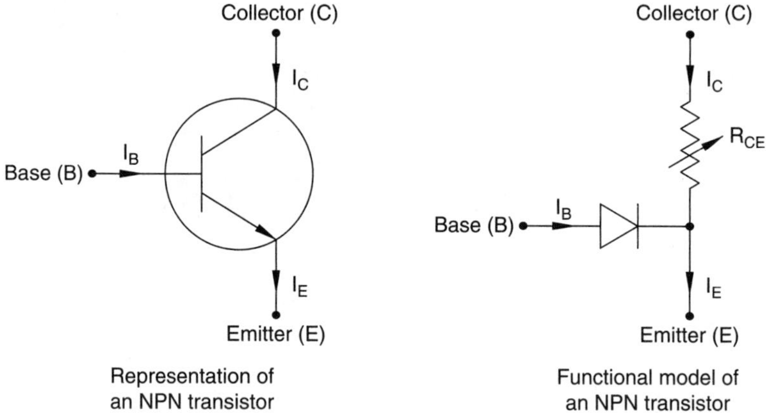

Collector (C)

I_C

I_B

Base (B)

I_E

Emitter (E)

Representation of
an NPN transistor

Collector (C)

I_C

R_{CE}

I_B

Base (B)

I_E

Emitter (E)

Functional model of
an NPN transistor

FIGURE 12.1   An NPN transistor.

The ratio of I_C to I_B is called the *current gain* of the transistor, a typical value being 100. (Since the current gain is typically very high, I_C and I_E would be nearly equal.) This describes the use of a transistor as a current amplifier. However, beyond a certain value of base current, R_{CE} becomes nearly zero. As a result, I_C does not increase any further, and the transistor is said to be *saturated*. In saturation mode, I_C depends on external resistance (load) and power supply, with the transistor offering nearly zero resistance to the current flow. Such an ideal transistor characteristic is shown in Fig. 12.2.

Thus, a transistor can be used as an electronic switch, in saturation mode. For such a use of a transistor, the circuit (regulating the base current) is designed in such a manner that the transistor remains either OFF ($I_B = 0$, resulting in $I_C = 0$) or becomes saturated.

The digital inputs/outputs of a PLC use similar electronic switching. However, to protect the internal circuits from external currents, generally some kind of *optoisolator* is used. An optoisolator is a device consisting of a light-producing element, such as an LED, and a light-sensing element, such as a phototransistor. When a voltage is applied

FIGURE 12.2   An
ideal transistor
characteristic.

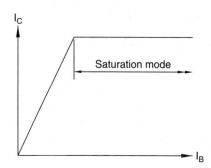

I_C

Saturation mode

I_B

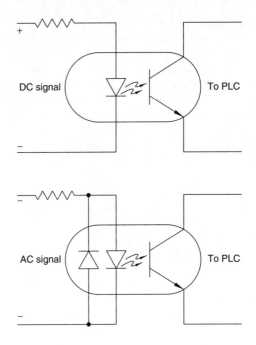

FIGURE 12.3
Typical PLC input
circuits.

to the LED, light is produced that is sensed by the photodetector. Sufficient light causes the phototransistor to saturate and it starts conducting. This arrangement completely isolates the PLC from external current/voltage, as the connection between the two is through a light beam only. For AC inputs, two opposing LEDs are used. Figure 12.3 shows this arrangement for both DC and AC input signals. DC signals can be given to AC inputs also, without bothering about the polarity. Hence, AC-type input is actually AC/DC-type input. A current-limiting resistor is needed to limit the current through the LED. The solid-state output terminals also generally use a similar isolation technique.

12.3   Input Types and Wiring

A commercial PLC provides for several types of inputs such as DC-, AC-, and analog-type "normal" inputs, apart from inputs for special functions such as high-speed input and immediate input. The PMC, however, usually accepts only 24-V DC-type normal inputs. A restriction is that an **external** power source must not be used for providing input signals; 24-V DC from inside the PMC remains available for this purpose, on a pin of each 50-pin connector on the input/output unit (on pin number B01 of CB104/105/106/107 on 0i control: 0 V is available on pin number A01). This is because the PMC generally does not have PLC-like optoisolators inside it. Note, however, that various types of inputs/outputs, with or without isolation, can be made available as options, by inserting certain input/output cards in the input/output unit of the PMC. The standard configuration usually accepts only 24-V DC sink/source inputs, and provides source-type outputs (designed for powering external devices, such as a relay, working on a 24-V DC **external** power supply). Our discussion in this chapter would remain limited to these input/output types only.

The input-signals (referred to as DI signals) to the PMC, through connectors CB104/105/106/107, are generally sink type (a type that drains energy). Hence, +24 V, as ON input signals, must be provided at respective terminals. Some DI signals, however, can be set to either sink type or source type, though a source-type DI signal is undesirable from the viewpoint of safety. This is because a source-type input requires that 0 V, as ON input signal, be provided to the input terminal. This is dangerous, because a grounded input line (due to a fault) would be interpreted as ON input signal. Hence, it is recommended that all DI signals be used as sink-type signals.

Figure 12.4 explains the method of providing sink-type DI signal (digital, of course) to the PMC. The wiring for source-type DI signal would be similar, with the difference that instead of +24 V, 0 V available at pin number A01 is used as ON input signal. DI signal X0004.0 through X0004.7 (which are provided through connector CB106) can be used both as source- and sink-type signals. COM4 terminal (terminal number A14) of CB106 is connected to 0 V (terminal number A01) when these signals are used as sink-type signals. For source-type applications, COM4 is connected to +24 V (terminal number B01). **COM4 must never be left open,** otherwise X0004.0 through X0004.7 would always remain OFF, irrespective of input signal condition. The external switch represents the discrete switching-action (ON/OFF) by an external switch/sensor.

12.4   Connector Pin Assignment

The block diagram of the general arrangement (on 0i control) for input/output connections is shown in Fig. 12.5. The PMC resides in the CNC control unit. It is connected to the I/O unit through an I/O link (which is a standard Fanuc cable). The I/O unit, terminal strip, and relay PCB can be seen in the MTB-designed electrical cabinet of the machine. On the I/O unit, four 50-pin connectors (CB104, CB105, CB106, and CB107) are present for receiving/sending input/output signals. These are connected to the terminal strip through four ribbon (flat) cables. Every pin of these connectors is used for some specific input/output signal, as given in Tables 12.1 and 12.2. Input wires from switches/sensors are directly connected to the terminal strip, but output devices are driven through relays, as these may require high current and/or AC supply.

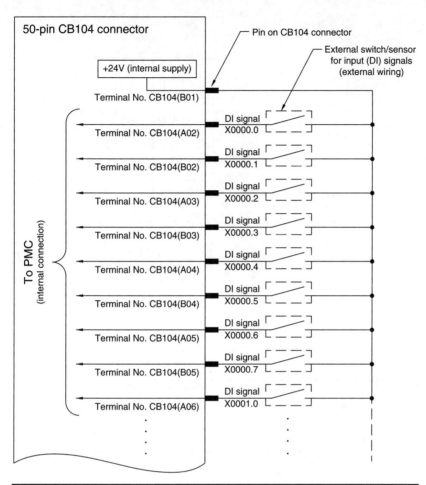

FIGURE 12.4   Input wiring for sink-type DI signals.

Typically, 96 inputs (plus 24 inputs for *manual pulse generators* and eight for DO-alarm detection) and 64 outputs are available. On 0i control, X0 to X11 (each consisting of eight bits, e.g., X0000.0 through X0000.7) are used for 96 DI signals, X12 is used for MPG, X13 and X14 (eight signals each) for additional MPGs (if any), and X15 (eight signals) for DO-alarm detection (e.g., overcurrent or abnormal temperature inside DO driver, caused by overload due to reasons such as accidental grounding of power cable). Y0 to Y7 (each consisting of eight signals, e.g., Y0000.0 through Y0000.7) are used for DO signals.

The 96 inputs and all 64 outputs can be used for any purpose. However, if the machine has Fanuc MOP, X4 to X11 are connected to various keys of MOP. Thus, only X0 to X3 (32 signals) are general-purpose DI signals, some of which might be used by the MTB, and

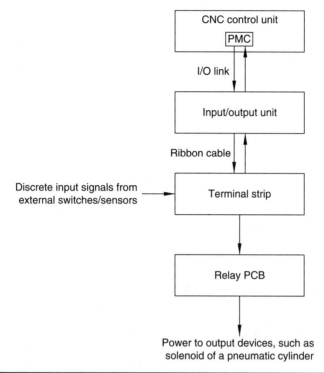

Figure 12.5   Block diagram of input/output connections with PMC.

CB104			CB105		
Pin No.	**Row A**	**Row B**	**Pin No.**	**Row A**	**Row B**
01	0V	+24V	01	0V	+24V
02	X0000.0	X0000.1	02	X0003.0	X0003.1
03	X0000.2	X0000.3	03	X0003.2	X0003.3
04	X0000.4	X0000.5	04	X0003.4	X0003.5
05	X0000.6	X0000.7	05	X0003.6	X0003.7
06	X0001.0	X0001.1	06	X0008.0	X0008.1
07	X0001.2	X0001.3	07	X0008.2	X0008.3
08	X0001.4	X0001.5	08	X0008.4	X0008.5
09	X0001.6	X0001.7	09	X0008.6	X0008.7
10	X0002.0	X0002.1	10	X0009.0	X0009.1
11	X0002.2	X0002.3	11	X0009.2	X0009.3
12	X0002.4	X0002.5	12	X0009.4	X0009.5

Table 12.1   Pin Assignment for CB104 and CB105 Connectors

CB104		
Pin No.	Row A	Row B
13	X0002.6	X0002.7
14		
15		
16	Y0000.0	Y0000.1
17	Y0000.2	Y0000.3
18	Y0000.4	Y0000.5
19	Y0000.6	Y0000.7
20	Y0001.0	Y0001.1
21	Y0001.2	Y0001.3
22	Y0001.4	Y0001.5
23	Y0001.6	Y0001.7
24	DOCOM	DOCOM
25	DOCOM	DOCOM

CB105		
Pin No.	Row A	Row B
13	X0009.6	X0009.7
14		
15		
16	Y0002.0	Y0002.1
17	Y0002.2	Y0002.3
18	Y0002.4	Y0002.5
19	Y0002.6	Y0002.7
20	Y0003.0	Y0003.1
21	Y0003.2	Y0003.3
22	Y0003.4	Y0003.5
23	Y0003.6	Y0003.7
24	DOCOM	DOCOM
25	DOCOM	DOCOM

TABLE **12.1**   (Continued)

CB106		
Pin No.	Row A	Row B
01	0V	+24V
02	X0004.0	X0004.1
03	X0004.2	X0004.3
04	X0004.4	X0004.5
05	X0004.6	X0004.7
06	X0005.0	X0005.1
07	X0005.2	X0005.3
08	X0005.4	X0005.5
09	X0005.6	X0005.7
10	X0006.0	X0006.1
11	X0006.2	X0006.3
12	X0006.4	X0006.5
13	X0006.6	X0006.7

CB107		
Pin No.	Row A	Row B
01	0V	+24V
02	X0007.0	X0007.1
03	X0007.2	X0007.3
04	X0007.4	X0007.5
05	X0007.6	X0007.7
06	X0010.0	X0010.1
07	X0010.2	X0010.3
08	X0010.4	X0010.5
09	X0010.6	X0010.7
10	X0011.0	X0011.1
11	X0011.2	X0011.3
12	X0011.4	X0011.5
13	X0011.6	X0011.7

TABLE **12.2**   Pin Assignment for CB106 and CB107 Connectors

CB106			CB107		
Pin No.	Row A	Row B	Pin No.	Row A	Row B
14	COM4		14		
15			15		
16	Y0004.0	Y0004.1	16	Y0006.0	Y0006.1
17	Y0004.2	Y0004.3	17	Y0006.2	Y0006.3
18	Y0004.4	Y0004.5	18	Y0006.4	Y0006.5
19	Y0004.6	Y0004.7	19	Y0006.6	Y0006.7
20	Y0005.0	Y0005.1	20	Y0007.0	Y0007.1
21	Y0005.2	Y0005.3	21	Y0007.2	Y0007.3
22	Y0005.4	Y0005.5	22	Y0007.4	Y0007.5
23	Y0005.6	Y0005.7	23	Y0007.6	Y0007.7
24	DOCOM	DOCOM	24	DOCOM	DOCOM
25	DOCOM	DOCOM	25	DOCOM	DOCOM

TABLE 12.2   Pin Assignment for CB106 and CB107 Connectors (*Continued*)

the rest remain free. Similarly, several of the output signals are connected to the LEDs of the Fanuc MOP. However, in many cases, the MTB designs its own MOP, and selects which signal to use for which purpose. The unused signals can be used for communicating with external devices. One would need to look into the electrical-interface manual supplied by the MTB to find out which signals have been used by them. The remaining signals are free, and can be used by the users.

If the number of DI/DO points is not sufficient, additional I/O units would need to be added. On the other hand, if even the minimum available points exceed the requirement, the MTB may not provide connection for all the four connectors. In such cases, the terminal strip typically does not have physical connection for CB107.

12.5   Discrete Sensors for Sourcing/Sinking PLC Inputs

In the standard PMC configuration, only discrete input/output signals are allowed. Input signals can be provided by mechanical switches (*momentary-* or *maintained-*type ON/OFF switch, or a limit switch) or proximity sensors. The proximity sensors have an electronic switching system inside them. These are typically *normally open* (N/O) type, and when an object comes within their designed range, the contact closes, simulating a switching action.

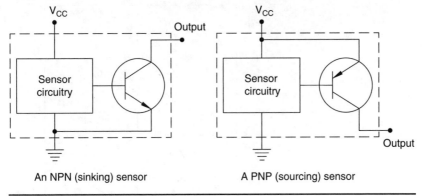

FIGURE **12.6** Discrete proximity sensors.

Since an internal transistor is typically used for switching, such sensors can be both PNP and NPN types. The PNP-type sensor supplies $+V_{CC}$ on its output terminal, whereas an NPN-type sensor connects it to ground (0 V), when switched ON. In the OFF state, the output terminal remains electrically open. Thus, a PNP sensor provides a sourcing output ($+V_{CC}$), and hence it is called a source-type sensor. Similarly, an NPN sensor is called a sink-type sensor (since it draws current through its output terminal). Figure 12.6 shows both types of sensors.

Electrical circuit can be complete only if the current supplied by a source is drained to a sink. **Therefore, a sourcing sensor can only be connected to a sinking input of PLC. Similarly, a sinking sensor is connected to a sourcing input.** While both possibilities exist, usually all PMC inputs are configured as sink-type inputs, so only source-type sensors are connected to these. Figure 12.7 shows how a source-type sensor is connected to a sink-type PLC input. Connection to the PMC is done in a similar manner. (The input of the PMC may not have optoisolation.)

A limitation of sink-type PLC inputs is that both the sensor and the PLC must use the **same** V_{CC}. A sourcing PLC input, on the other hand, allows the sensor to work on a **different** V_{CC} also; only the *ground* should be common. This is useful when the operating-voltage requirements of the sensor and the PLC are different. Such an arrangement is shown in Fig. 12.8, where a sink-type sensor is connected to a source-type PLC input. Note that the inputs of the PLC shown in this figure are actually source/sink type (AC type) that can be wired as both source- and sink-type. Here, source-type wiring is shown, which is needed for a sink-type sensor. The operating voltages of the PLC and the sensor, V_{CC1} and V_{CC2}, respectively, can be the same or different. However, as discussed earlier, this type of wiring is not recommended for safety considerations.

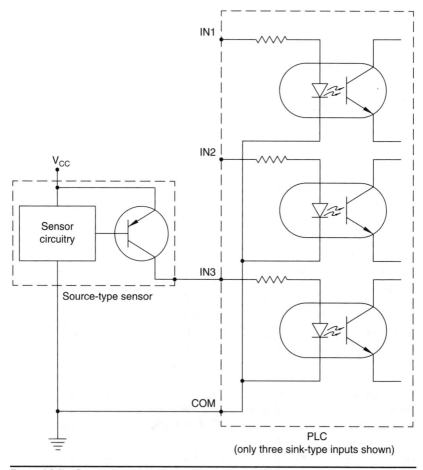

FIGURE **12.7** Sourcing sensor connected to a sinking PLC input.

12.6 Output Types and Wiring

In commercial PLCs, several types of outputs are available, employ-
ing different technologies. The PMC used in 0i control, however, uses
N-channel enhancement-type MOSFET switching technique. The
switching action is similar to that of an NPN transistor, where *source*,
gate, and *drain* of MOSFET correspond, respectively, to emitter, base,
and collector of NPN transistor.

A major difference between a MOSFET and a transistor is that
MOSFET is voltage-controlled, whereas transistor is current-controlled
(for required biasing). When a voltage is applied at the gate, an electro-
static field is generated which creates a conducting channel (P-type or
N-type) between drain and source. No current flows through the

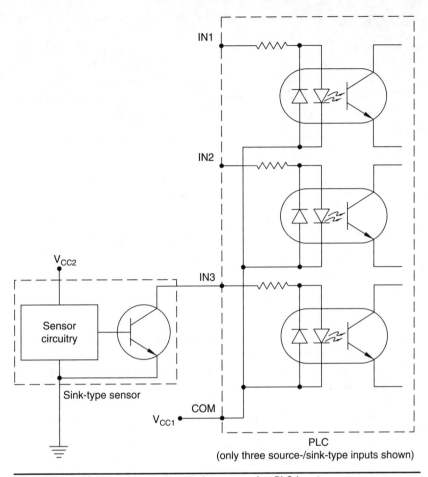

FIGURE 12.8 Sinking sensor connected to a sourcing PLC input.

gate. This not only can provide isolation between input and output circuits, but the input load requirement also remains negligible, which is the main advantage of a MOSFET over transistors.

The conductivity of the conducting channel between drain and source of a MOSFET depends on the voltage level on the gate. As in case of switching using transistors, MOSFET also is used for switching in its saturated mode of operation (i.e., the drain current is either zero or remains constant at its maximum value, for a given gate voltage). Figure 12.9 explains the switching action of an N-channel MOSFET. (The other type is P-channel MOSFET. The difference between the two is at the construction level. Overall circuit connection is same, except that polarity of source and drain changes. Its switching action is similar to that of a PNP transistor.) When the switch is pressed, the LED lights up. Notice that the high resistance connected to the gate

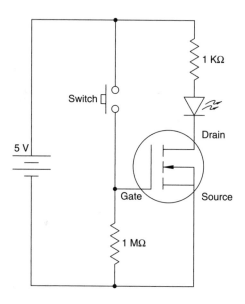

Figure 12.9 An N-channel enhancement-type MOSFET as a switch.

consumes negligible power. Refer to a basic text on solid-state devices for more information about MOSFETs.

The output signals of PMC are designed to be used as **sourcing** outputs, with MOSFETs used as switching devices. A typical output wiring for CB104 (wiring for CB105, CB106, and CB107 are similar; refer to Tables 12.1 and 12.2 for pin connections) is shown in Fig. 12.10, where external relays, powered by external 24-V DC power supply, are being controlled by DO signals. Relays are used because a MOSFET cannot take very heavy current. Moreover, it cannot be used for AC devices. A relay obviates such problems.

All output terminals must be wired **independently**, as shown in the figure. Never short two (or more) output terminals to drive a single device. Also note the presence of reverse-biased diodes, connected in parallel with each relay coil. Such a diode prevents excessive voltage built-up across the coil of the relay, when the MOSFET signal turns OFF. If unchecked, the developed voltage can rise up to several hundred volts, which can damage the MOSFET. The diode shunts the induced current, preventing voltage rise-up. In fact, some relays are manufactured with this diode preinstalled.

A limitation of source-type outputs is that all output devices would have to work on the same V_{cc} (though this does not matter because we use only relays as output devices, and we can always decide to use same type of relays), since V_{cc} is connected to the common output-terminal (DOCOM). In a sink-type output (which is not available with PMC), 0 V is connected to the common output-terminal, and V_{cc} goes directly to output devices. Hence, it is possible to use

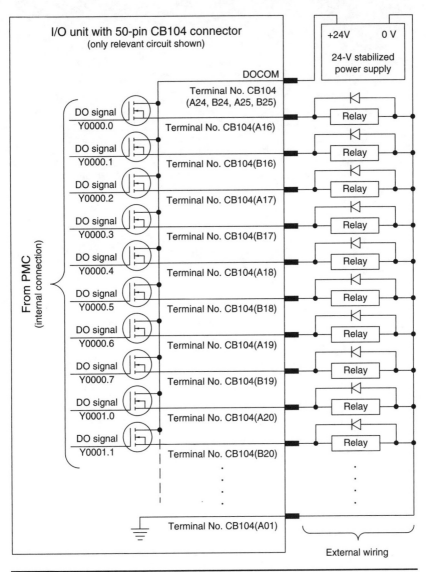

FIGURE 12.10 Output wiring for source-type DO signals.

different V_{CC} with different devices. This arrangement, however, has a serious drawback. If, due to some fault, the wire (on the PLC side) of an output device gets grounded, the device would switch ON without any output signal. Hence, from safety considerations, source-type outputs are preferred over sink-type outputs. Of course, with PMC, there is no option other than source-type outputs.

CHAPTER **13**

Programmable
Data Input

13.1   Introduction

This feature of a CNC machine is not really related to macro programming. It has been included in this text because it is an advanced
programming feature which many programmers are not aware of,
even though it is very useful and its use is pretty straightforward.

As the name suggests, *programmable data input*, which is commanded by G10, is used to write certain data into the control such as
offset values and parameters. G10 is an optional feature on some control versions, which would need to be activated if its use is desired.
This feature does not allow reading the current values of such data;
one can only overwrite these.

The main applications of G10 are data input for the following:

- Offset distances for workpiece coordinate systems
- Offset distances for additional workpiece coordinate systems
  on a milling machine
- Compensation values on a milling machine (length and
  diameter)
- Compensation values on a lathe (radial/axial values, nose
  radius, and tip number)
- Parameter values

G10 can also be used to feed tool-management data such as tool
life and offset number. This is, however, not discussed in this chapter,
since it is not a very common application.

If the macro-programming feature is available on the machine,
G10 is not needed in most cases. This is because system variables for
all the offset distances and compensation values are available as read/
write variables (refer to Sec. 3.5). In fact, system variables offer more
flexibility, since the current values stored in them can also be read

which is not possible with G10. System variables are available even for some of the parameters (though very few of them are covered). However, for the remaining parameters and tool-life data, G10 has to be used, if their values are to be altered through a program.

G10 can be used in both absolute as well as incremental modes, for specifying offset distances and compensation values. In absolute mode, the specified values simply overwrite the existing values; in incremental mode, the specified values are added (with sign) to the existing values.

The effect of G10 is permanent, that is, it remains valid for all the subsequent machining sessions, until the values are changed again by G10 or otherwise.

13.2 Data Input for WCS Offset Distances

Offset distances for external and G54 through G59 workpiece coordinate systems can be specified for each axis, in absolute or incremental mode. The syntax is the same for both lathe and milling machines:

```
G10 L2 P_ X_ Y_ Z_;
```

where

        L2 selects this category of data input,
        P0 refers to external WCS,
        P1 refers to G54 WCS,
        P2 refers to G55 WCS,
        P3 refers to G56 WCS,
        P4 refers to G57 WCS,
        P5 refers to G58 WCS,
        P6 refers to G59 WCS,
        X, Y, and Z contain the desired values for the respective axes.

On a milling machine, G90/G91 would need to be commanded before G10, for selecting absolute/incremental modes. On a lathe, with G-code system A, X/Z are used for absolute values, and U/W for incremental values (the Y-axis is generally not available on a lathe).

Incremental mode is very useful for correcting the offset distances by the observed error. For example, on a milling machine, if it is desired to shift the G54 WCS by 0.1 mm in the negative Z-direction (so that the same tool now machines at a level of 0.1 mm lower than the previous level, without modifying the program), one may command (in MDI mode, if the adjustment is to be done only once)

```
G91 G10 L2 P1 Z-0.1;
```

Of course, this can also be done through the associated system variable (refer to Table 3.12):

```
#5223 = #5223 - 0.1;
```

Both methods have the same effect. Selection between the two is a matter of individual choice. However, if the macro option is not enabled on the machine, G10 has to be used. In fact, G10 also is an optional feature, but now it is usually available with the standard control package.

13.3   Data Input for Additional WCS Offset Distances

Besides the six standard WCS, selectable with G54 through G59, 48 additional WCS are optionally available on milling machines (in fact, this option is available for up to 300 additional WCS also). These are invoked in a program by

```
G54.1 P<n>;
```

or simply by

```
G54 P<n>;
```

where n = 1 to 48.

These are referred to as first additional WCS, second additional WCS, and so on, corresponding to n = 1, 2, ... A P-word must be specified with G54.1/G54. If it is not specified, P1 (implying first additional WCS) is automatically assumed by the control.

Offset distances for these can be specified in a manner similar to that used in Sec. 13.2:

```
G10 L20 P_ X_ Y_ Z_;
```

where

> L20 selects this category of data input,
> P1 refers to first additional WCS,
> P2 refers to second additional WCS,
> . . .
> . . .
> P48 refers to 48th additional WCS,
> X, Y, and Z contain the desired values for the respective axes.

Both absolute and incremental modes can be used. As an example, the following command would shift the first additional WCS by 0.1 mm in the positive Z-direction:

```
G91 G10 L20 P1 Z0.1;
```

This, of course, can also be done through system variables (refer to Table 3.12):

```
#7003 = #7003 + 0.1;
```

13.4   Data Input for Compensation Values on a Milling Machine

The description given here refers to C-type compensation memory (also referred to as C-type offset memory), used in 0i series controls. Compensation values for length (H-code) and radius (D-code) of a milling cutter can be specified as described in Table 13.1, where the P-word defines the offset number (typically, 1 through 400) and the R-word contains the corresponding compensation value. To provide compatibility with programs written for older control models, the system allows L1 to be specified instead of L11. The permissible input range for compensation values, in IS-B (Increment System B), is given in Table 13.2. The valid range is much larger than what might actually be needed.

Some examples are given below. Note that the given descriptions are valid only when these commands are executed sequentially:

G90 G10 L12 P10 R5; (Enters 5 mm into offset number 10, as the geometry compensation value for D-code, making the tool radius equal to 5 mm, assuming that the wear value for the radius is zero. Note that the D-code refers to the **radius**, not to the diameter)

G91 G10 L12 P10 R5; (Enters 5 mm incrementally into offset number 10, as the geometry compensation value for the D-code, making the tool radius equal to 10 mm, which was 5 mm earlier)

G90 G10 L13 P10 R−0.5; (Enters −0.5 mm into offset number 10, as the wear compensation value for the D-code, making the tool radius equal to 9.5 mm)

Compensation Value	Format
Geometry compensation value for H-code	G10 L10 P_ R_;
Wear compensation value for H-code	G10 L11 P_ R_;
Geometry compensation value for D-code	G10 L12 P_ R_;
Wear compensation value for D-code	G10 L13 P_ R_;

TABLE 13.1   Programmable-Data-Entry Format for H- and D-Code Values

Geometry Compensation Value		Wear Compensation Value	
Metric input	Inch input	Metric input	Inch input
± 999.999 mm	± 99.9999 inch	± 99.999 mm	± 9.9999 inch

TABLE 13.2   The Valid Input Range for Compensation Values in IS-B

G91 G10 L13 P10 R-0.05;   (Enters –0.05 mm incrementally into off-set number 10, as the wear compensation value for D-code, changing its value to –0.55, which makes the tool radius equal to 9.45 mm)

G90 G10 L10 P10 R-500;   (Enters –500 mm into offset number 10, as the geometry compensation value for H-code)

G91 G10 L10 P10 R5;   (Enters 5 mm incrementally into offset number 10, as the geometry compensation value for H-code, making it smaller by 5 mm, to become –495 mm. So, the same tool would dig 5 mm less into the workpiece, for the same program, if G43 is being used for length compensation)

G91 G10 L11 P10 R0.5;   (Enters 0.5 mm incrementally into offset number 10, as the wear compensation value for H-code. So, the wear value would increase by 0.5 mm, and hence, the same tool would dig 0.5 mm less into the workpiece, for the same program, if G43 is being used)

Of course, system variables also can be used for the same functionality. For example, the first and the last examples are, respectively, equivalent to (refer to Table 3.4)

```
#13010 = 5;
#10010 = #10010 + 0.5;
```

13.5   Data Input for Compensation Values on a Lathe

Geometry as well as wear values for radial/axial distances, nose radius, and tip number can be specified. A difference from the previously described formats is that no L-word is used for entering compensation values on a lathe. The format is

G10 P_ X_ Y_ Z_ R_ Q_;   (Absolute mode)
G10 P_ U_ V_ W_ C_ Q_;   (Incremental mode)

where
P0 refers to work-shift values,
P1 through P64 refer to wear compensation values corresponding to offset numbers 1 through 64 (assuming availability of 64 offset numbers),
P10001 through P10064 refer to geometry compensation values corresponding to offset numbers 1 through 64,
X/U, Y/V (if the Y-axis is available), and Z/W contain the desired values for the respective axes in absolute/incremental mode,

R/C contain nose radius in absolute/incremental mode, and Q contains tip number (in both modes). The tip number or nose number has been arbitrarily defined by Fanuc for different tool orientations. Figure 13.1 shows Fanuc-defined tip numbers on a rear-type lathe. Note that the value of Q, specified for a geometry offset number, automatically assigns the same value to the corresponding wear offset number, and vice versa. In other words, the geometry and wear compensation values for Q are always same. This is a built-in safety feature of the control which eliminates the possibility of inadvertent error is specifying the tip number.

Not all the compensation values in the G10 block need be specified. The unspecified values remain unchanged. Some examples are given below:

`G10 P10010 X-100 W5;`	(Corresponding to geometry offset number 10, i.e., in the G10 row on the geometry offset screen, the radial compensation value is set to −100 mm, and the axial compensation value is increased by 5 mm. Mixed coordinate mode is allowed in G10)
`G10 P10 X0 Z0;`	(Corresponding to wear offset number 10, the radial as well as the axial compensation values are set to zero. In other words, the radial and axial wear offsets are made zero, for the W10 row on the wear offset screen)
`G10 P10010 R0.8 Q3;`	(Corresponding to geometry offset number 10, i.e., in the G10 row on the geometry offset screen, the nose radius is set to 0.8 mm. The tip number is set to 3 on both geometry and wear offset screens, corresponding to offset number 10)

In this case also, it is possible to use system variables to have the same result. The three G10 blocks given above can be replaced by the following blocks (refer to Table 3.2):

```
For G10 P10010 X-100 W5; use
    #2710 = -100;
    #2810 = #2810 + 5;
For G10 P10 X0 Z0; use
    #2010 = 0;
    #2110 = 0;
For G10 P10010 R0.8 Q3; use
    #2910 = 0.8;
    #2310 = 3;
```

As another example, if it is desired to clear all the 64 offsets, G10 can be executed in a loop. If 99 offsets are available, just replace 64 by 99 in the given program:

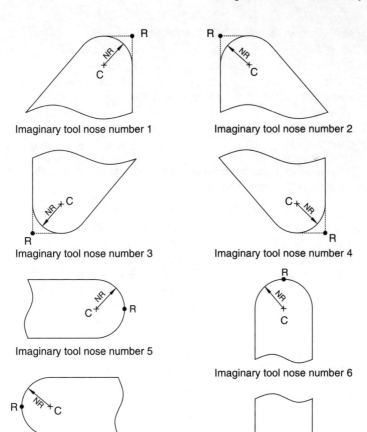

Imaginary tool nose number 1

Imaginary tool nose number 2

Imaginary tool nose number 3

Imaginary tool nose number 4

Imaginary tool nose number 5

Imaginary tool nose number 6

Imaginary tool nose number 7

Imaginary tool nose number 8

R : Reference point of the tool
C : Center of the tool nose
NR: Nose radius

Note:

1. The usual method of offset setting establishes the imaginary point R as the reference point of the tool. There are eight possible orientations of this point with respect to the nose center C, depending on the type of the tool (left-hand, right-hand, or neutral) and its orientation with respect to the workpiece. Each orientation has been assigned a unique tool nose number, 1 through 8.

2. It is also possible to make the nose center C the reference point of the tool. This is done by adding the nose radius to the Z-value, and twice the nose radius to the X-value, while measuring the offset distances. In such a case, the assigned tool-nose number is 0. A serious drawback of this method is that this type of offset setting cannot be used for machining without radius compensation because of excessive error, even in straight turning and straight facing. The usual method of offset setting ensures that at least in straight turning and straight facing, there would be no error due to nose radius, even if radius compensation is not used.

Figure 13.1 Fanuc-defined tip numbers for different tool orientations on a rear-type lathe.

```
O8024 (CLEAR LATHE OFFSETS WITH G10);
#100 = 1;                        (Counter initialized)
WHILE [#100 LE 64] DO 1;         (Start of loop)
G10 P#100 X0 Z0 R0 Q0;           (Clears wear offset)
G10 P[10000 + #100] X0 Z0 R0 Q0; (Clears geometry offset)
#100 = #100 + 1;                 (Counter incremented)
END 1;                           (End of loop)
M99;                             (Returns to the calling program)
```

The same thing can be done using system variables also, though the G10 method is simpler in this case:

```
O8025 (CLEAR LATHE OFFSETS WITH SYSVAR);
#100 = 1;                 (Counter initialized)
WHILE [#100 LE 64] DO 1;  (Start of loop)
[#2000 + #100] = 0;       (Clears X-axis wear offset)
[#2100 + #100] = 0;       (Clears Z-axis wear offset)
[#2200 + #100] = 0;       (Clears nose radius wear offset)
[#2300 + #100] = 0;       (Clears tip number)
[#2700 + #100] = 0;       (Clears X-axis geometry offset)
[#2800 + #100] = 0;       (Clears Z-axis geometry offset)
[#2900 + #100] = 0;       (Clears nose radius geometry offset)
#100 = #100 + 1;          (Counter incremented)
END 1;                    (End of loop)
M99;                      (Returns to the calling program)
```

13.6   Data Input for Parameter Values

By now, the readers might have gotten an impression that there is no need to use G10 if the macro-programming option is available on the machine. The system-variable method, however, is limited only to whatever has been discussed so far. Most of the parameters do not have associated system variables. Therefore, G10 is the only method to change parameters through a program. Such a use of G10 is often referred to as *programmable parameter entry*.

A difference from the previous applications of G10 is that it behaves as a modal code, when used for parameter entry. Once it enters parameter-entry mode, as many parameters as desired can be entered in subsequent blocks, until this mode is canceled by G11. Another difference is that no decimal values are allowed. Therefore, if the parameter relates to some axis distance, the value must be specified in the least input increment (i.e., in microns in millimeter mode, and in steps of 0.0001 in. in inch mode). Note that if the same parameter also has a system variable defined for it, millimeter or inch (as appropriate) value is used in such a variable.

Note that the parameter values that are most suitable for most of the cases are already set by the MTB. **Any change in these values is not recommended because an incorrect setting may result in unpredicted, possibly dangerous, behavior of the machine, causing injury to the operator or damage to the machine or both.** However, certain parameters, such as those for software overtravel limits, G-code system selection (A, B, or C) on a lathe, straight-line positioning with G00, calculator-type data entry (so that X10, e.g., is interpreted as 10 mm, not as 10 μm), etc. may need to be changed to suit a particular application. However, in such cases, first note down the original settings before making any change, so that one may revert back to the original values, in case of any incorrect setting. Any change must be done by an experienced person, after thoroughly studying the parameter which is intended to be changed. The complete description of all the parameters can be found in the *Parameter Manual* supplied with the machine.

Further discussion necessitates a clear understanding of types of parameters. There are, broadly speaking, four types of parameters: *bit type* (permissible data 0 or 1), *byte type* (permissible data range −128 to 127, or 0 to 255), *word type* (permissible data range −32,768 to 32,767, or 0 to 65,535) and *two-word type* (permissible data range −99,999,999 to 99,999,999). The bit-type parameters have rows of eight bits (bit #0 at the extreme right and bit #7 at the extreme left). Some bit-type parameters have multiple rows of eight bits in each row, corresponding to each machine axis. Several other parameter-types also have multiple rows for different axes. On a two-axis lathe, the first row corresponds to the X-axis, and the second row to the Z-axis. On a three-axis milling machine, the three rows correspond to the X-, Y-, and Z-axes, respectively. Such parameters are called *axis-type* parameters. For the purpose of effective memory utilization, different types of parameters are used for different applications. For example, I/O CHANNEL is a byte-type parameter, whereas the parameters for setting software overtravel limits (distances in microns, in machine coordinate system) are two-word parameters.

The format for programmable parameter entry is as follows:

```
G10 L50;      (Parameter entry mode starts)
N_ R_;        (For parameters other than axis-type)
N_ P_ R_;     (For axis-type parameters)
. . .
. . .         (Multiple parameter entry is permitted)
. . .
G11;          (Parameter-entry mode ends)
```

where

L50 selects parameter-entry mode,
N contains parameter number,
P contains axis number (used for axis-type parameters only),
R contains the specified value of the parameter.

NOTE:

1. In a bit-type parameter, it is not possible to change a single bit independently. All eight bits must be specified.

2. A decimal cannot be used in the value specified in the R-word.

3. Other NC statements cannot be specified between G10 L50 and G11.

4. If a canned cycle is being used, it must be canceled before commanding G10 L50.

5. For axis-type parameters, P1 refers to the first row, P2 to the second row, and so on.

6. Inserting sequence numbers in the given format is permitted. If two N-words appear in a block, the one at the left is considered the sequence number, and the other parameter number.

Example 1 (Word-Type Parameter):
G-code that calls custom-macro-program number O9010 is stored in parameter 6050, on both lathe and milling machine. Therefore, if it is desired to call O9010 by, say, G100 (any value from 1 to 9999 can be selected; note that the acceptable range of a word-type parameter is larger than the defined range for parameter 6050), 100 would need to be stored in parameter 6050. This can be done by

```
G10 L50;
N6050 R100;
G11;
```

Example 2 (Two-Word Axis-Type Parameter):
It is desired to set external offset to zero, G54 X-offset to 150.123 mm, and G54 Z-offset to 300.456 mm on a lathe. The associated parameters are 1220 and 1221, respectively. The values are required to be specified in microns. The first row of such parameters corresponds to the X-axis, and the second to the Z-axis, on a lathe. On a milling machine, the third row would correspond to the Z-axis. The desired changes can be done by

```
G10 L50;
N1220 P1 R0;
N1220 P2 R0;
N1221 P1 R150123;
N1221 P2 R300456;
G11;
```

Of course, the G10 L2 method also can be used (which is a bit simpler method), which does not directly refer to these parameters:

```
G10 L2 P0 X0 Z0;
G10 L2 P1 X150.123 Z300.456;
```

Note that the distances are specified in millimeters, not in microns. A flexibility in G10 L2 method is that changes can be done in both absolute and incremental modes. G10 L50 method does not permit incremental changes; R-word always contains the **new** value. The system-variable method can also be used in this case.

Example 3 (Byte-Type Parameter):
If a compact PCMCIA card is desired to be used as an external memory device, parameter 0020 would need to store 4, on both lathe and milling machines. The permissible range of this parameter is 0 to 35 (which is smaller than the defined range for a byte-type parameter). The desired change can be done by

```
G10 L50;
N20 R4;
G11;
```

When specifying parameter number, leading zeroes can be omitted. Therefore, N20 and N0020 are both equivalent in this example.

Example 4 (Bit-Type Parameter):
It is desired to use calculator-type decimal input for distances (so that, e.g., X10 is interpreted as 10 mm, not as 10 μm). Parameter 3401#0 (which means bit #0, the one at the extreme right, of parameter 3401) is required to be set as 1 for this. However, the G10 L50 method requires that all the eight bits be specified, even if change in just one bit is needed. Let us assume that the current setting of parameter 3401 on a lathe is 00000010 (one has to find it out; there is no way out). This would need to be changed to 00000011, which can be done by

```
G10 L50;
N3401 R00000011;
G11;
```

Note that leading zeroes in the R-word **cannot** be omitted.

Example 5 (Bit-Axis-Type Parameter):
G51/G51.1 is available on a milling machine for mirroring of a selected toolpath. If, however, it is desired to **always** have mirroring for, say, the X-axis, this can be more conveniently done by setting parameter 0012#0 to 1 for the X-axis, which would obviate the need for using G51/G51.1 for mirroring. Let us assume that the current setting of parameter 0012 is 00000000, in the first row (i.e., for X-axis). This would need to be changed to 00000001, for the X-axis mirroring. The change can be done by

```
G10 L50;
N12 P1 R00000001;
G11;
```

Another, and a simpler, way to do this is to use system variable for mirroring, #3007 (see Table 3.9):

```
#3007 = 1;
```

A change in parameter 0012#0 automatically changes variable #3007, and vice versa.

In general, if a system variable corresponding to a parameter is available, there is no need to use G10, since the system-variable method would be simpler. Moreover, by reading system variables, it is also possible to know the current settings (to which one can revert back, if so desired), which is not possible with parameters which can only be overwritten with new values (if change through a program is desired).

Actually, G10 is an older feature that came before macro programming was introduced. Now, it has lost its relevance in many cases.

A Limitation of G10 L50 Method

Once a parameter is changed, there is no way it could be reverted back to its **original** value, unless we know its prechange value. (Then we can use G10 L50 again with the original values.) G10 L50 only overwrites a parameter with a new value(s). In fact, this is true for all applications of G10. It is not possible to "undo" a G10. While incorrect changes in other applications may not have serious implications, improper parameter setting may prove to be extremely dangerous. Therefore, extreme care must be exercised while using G10 L50. In fact, as a rule of thumb, one should not play with parameters unless it is absolutely necessary, and one fully understands the function of the parameter which is going to be changed. To be on the safer side, **one must always keep a backup of the original settings of all the parameters**. If you do not have an external memory device (such as a compact flash card) or RS-232 connection on your machine, or you simply do not know how to do the backup, just note down all the parameter values on a piece of paper and keep it in a safe place. An hour or two spent today may one day save weeks of downtime on the machine!

List of Complex Macros

List of Parameters

Parameter Number	Purpose	Page Number
0012#0	X-axis mirroring	261
0020	Input/output channel selection	148, 261
0102	Selection of I/O device such as Fanuc Handy File when 0020 is set to 0	148
1220	External offsets	260
1221	G54 offsets	260
1300 series	Software overtravel limits	107
1401#1	Rapid positioning in a straight line, i.e., without dog-leg effect	133
1827	In-position widths along different axes during feed motion (valid when 1801#4 = 1)	163
3202#0	Edit-protection of programs in the range 08000 – 08999	83, 138
3202#4	Edit-protection of programs in the range 09000 – 09999	83, 138
3202#6	Display of program numbers of edit-protected programs in directory search	84
3203#7	Retaining MDI program even after control reset	25
3204#0	Selection between square bracket and parenthesis on a small MDI keyboard	14
3204#2	Display of soft keys for "(", "),” and “@” on a small MDI keyboard	15
3204#6	Retaining MDI program after execution is over	24

Parameter Number	Purpose	Page Number
3401#0	Calculator-type decimal number interpretation	15, 22, 261
3404#6	Missing M30/M02 at program end	115
3411 – 3420	M-codes for preventing buffering	163, 166
3451#2	Spindle speed specification up to one decimal point on a milling machine	20
6000#3	Interchanging wear and geometry system variable numbers on a milling machine	44
6000#5	Single-block execution of macro statements	17
6001#5	Subprogram call using a T-code	150
6001#6	Retaining common variables even after control reset	29
6001#7	Retaining local variables even after control reset	29
6004#0	Solution range of ATAN	61, 233
6004#1	Normalization of SIN/COS/TAN values to zero, for angles close to 0°/90°	63
6006#0	Availability of AND/OR/XOR as Boolean functions in logical statements	73, 82
6006#1	Modal information up to previous/current block on a milling machine	52
6030	M-code for subprogram call from an external input/output device (such as M198)	148
6050 – 6059	G-codes for macro calls	146, 156, 214
6071 – 6079	M-codes for subprogram calls	148
6080 – 6089	M-codes for macro calls	147
6700#0	M-codes for part count increment	49
6710	Additional M-code for part count increment	49
6711	Number of parts produced	49
6712	Total number of parts produced	49
6713	Number of parts required to be produced	49

Index